616.9792 N9

Nutrition a....

DO679409

NUTRITION
and
AIDS

NUTRITION
and
AIDS

Edited by

Ronald R. Watson

CRC Press

Boca Raton   Ann Arbor   London   Tokyo

616. 9792
N959

Library of Congress Cataloging-in-Publication Data
Nutrition and AIDS / edited by Ronald Watson.
    p. cm. — (Modern nutrition series)
  Includes bibliographical references and index.
  ISBN 0-8493-7842-7
  1. AIDS (Disease)—Nutritional aspects. I. Watson, Ronald R.
(Ronald Ross) II. Series.
RC607.A26N885   1994
616.97'92-dc20   93-29784
                                                            CIP

This book contains information obtained from authentic and highly regarded sources. Reprinted material is quoted with permission, and sources are indicated. A wide variety of references are listed. Reasonable efforts have been made to publish reliable data and information, but the author and the publisher cannot assume responsibility for the validity of all materials or for the consequences of their use.

Neither this book nor any part may be reproduced or transmitted in any form or by any means, electronic or mechanical, including photocopying, microfilming, and recording, or by any information storage or retrieval system, without prior permission in writing from the publisher.

All rights reserved. Authorization to photocopy items for internal or personal use, or the personal or internal use of specific clients, may be granted by CRC Press, Inc., provided that $.50 per page photocopied is paid directly to Copyright Clearance Center, 27 Congress Street, Salem, MA 01970 USA. The fee code for users of the Transactional Reporting Service is ISBN 0-8493-7842-7/94/$0.00+$.50. The fee is subject to change without notice. For organizations that have been granted a photocopy license by the CCC, a separate system of payment has been arranged.

CRC Press, Inc.'s consent does not extend to copying for general distribution, for promotion, for creating new works, or for resale. Specific permission must be obtained in writing from CRC Press for such copying.

Direct all inquiries to CRC Press, Inc., 2000 Corporate Blvd., N.W., Boca Raton, Florida 33431.

© 1994 by CRC Press, Inc.

No claim to original U.S. Government works
International Standard Book Number 0-8493-7842-7
Library of Congress Card Number 93-29784
Printed in the United States of America 1 2 3 4 5 6 7 8 9 0
Printed on acid-free paper

SERIES PREFACE

The CRC Series in Modern Nutrition is dedicated to providing the widest possible coverage to topics in nutrition, in all its diversity, both basic and applied.

Published for an advanced or scholarly audience, the volumes of the Series in Modern Nutrition are designed to explain, review, and explore present knowledge and recent trends, developments, and advances across the field. As such, they will also appeal to the dedicated layman. Volumes in this series will reflect the broad scope of information available as well as the interests of authors. The diversity of topics will appeal to an equally diverse readership from volume to volume. The format for the series will vary with the needs of the author and the topic, including but not limited to, edited volumes, monographs, handbooks, and texts.

Contributors from any bona fide area of nutrition, including the controversial, are welcome.

James F. Hickson, Jr., Ph.D., R.D.
Ira Wolinsky, Ph.D.
Series Editors

CRC SERIES IN
MODERN NUTRITION

PREFACE

There is a wide variety of retroviruses infecting a broad spectrum of animals, including HIV-1 which infects man. In man, progression to disease, ARC or AIDS, seems variable. Development of pathogenesis is hypothesized to be affected by a variety of agents and conditions including age and immune status of the patient, other concurrent infections, and diet which modify immune functions. Yet it is not clear why there are largely differences in the rate that adults progress to disease after infection, which may take decades in some cases. The fact that various cofactors of agents are hypothesized to affect this process offers opportunities for study and lifestyle change which can influence progression to AIDS.

Cofactors may also affect susceptibility to the initial infection with retroviruses including HIV-1. Clearly, immunosuppressive conditions like protein malnutrition reduce resistance to some pathogens, while high intakes of vitamins stimulate immune functions which can increase resistance to infection. Because of cofactors such as these, other infections may influence the initial infection with HIV-1 as well as its progression rate to major pathogenesis. Such cofactors offer avenues to attack retroviral diseases and are tools to understand the mechanisms of action. As we develop greater knowledge of how cofactors modify the physiology and immunology of HIV-1 infected individuals, we will better understand retroviruses and their role in immunosuppression.

Ronald Ross Watson

THE EDITOR

Ronald Ross Watson, Ph.D., initiated and directed the Specialized Alcohol Research Center at the University of Arizona College of Medicine until 1994. A theme of this National Institute of Alcohol Abuse and Alcoholism (NIAAA) Center Grant is to understand the role of ethanol-induced immunosuppression on AIDS, including the use of nutrition to normalize immune dysfunction.

Dr. Watson attended the University of Idaho, but graduated from Brigham Young University in Provo, Utah with a degree in Chemistry in 1966. He completed his Ph.D. degree in 1971 in Biochemistry from Michigan State University. His postdoctoral schooling was completed at the Harvard School of Public Health in Nutrition and Microbiology, including a 2-year postdoctoral research experience in Immunology. He was an Assistant Professor of Immunology and did research at the University of Mississippi Medical Center in Jackson from 1973 to 1974. He was an Assistant Professor of Microbiology and Immunology at the Indiana University Medical School from 1974 to 1978 and an Associate Professor at Purdue University in the Department of Food and Nutrition from 1978 to 1982. In 1982 he joined the faculty at the University of Arizona in the Department of Family and Community Medicine, Nutrition Section, and is a Research Professor. He has published 300 research papers and review chapters.

Dr. Watson is a member of several national and international nutrition, immunology, and AIDS research societies. He directed the first symposium on nutrition and AIDS at FASEB, as well as one at the International Congress on Nutrition in Australia in 1993.

CONTRIBUTORS

Jeffrey M. Aron, M.D.
Division of Gastroenterology
Department of Medicine
Mt. Zion Medical Center
University of California,
  San Francisco
San Francisco, California

Marianna K. Baum, Ph.D.
Chief, Nutrition Division
Department of Epidemiology
  and Public Health
University of Miami
School of Medicine
Miami, Florida

**Eduardo A. Castellon-Vogel,
M.H.S.**
Department of Nutrition
University of North Florida
Jacksonville, Florida

Lawrence A. Cone, M.D.
Department of Medicine
Section of Infectious Diseases
Eisenhower Medical Center
Rancho Mirage, California

Gregg O. Coodley, M.D.
Division of Internal Medicine
Oregon Health Science University
Portland, Oregon

Nancy Cure, M.D.
Fogarty Scholar
Department of Epidemiology and
  Public Health
University of Miami
School of Medicine
Miami, Florida

Cyril O. Enwonwu, Ph.D.
Department of Biochemistry
University of Maryland
  at Baltimore
Baltimore, Maryland

Jacquelyn H. Flaskerud, Ph.D.
School of Nursing
University of California
Los Angeles, California

Silvia Franceschi, Ph.D.
Epidemiology Unit
Aviano Cancer Center
Aviano, Italy

Steve Harakeh, Ph.D.
Research Scientist
Viral Carcinogenesis
  and Immunology Laboratories
Linus Pauling Institute
  of Science and Medicine
Palo Alto, California

Raxit J. Jariwalla, Ph.D.
Senior Scientist and Head
Viral Carcinogenesis and
  Immunology Laboratories
Linus Pauling Institute
  of Science and Medicine
Palo Alto, California

Carlo La Vecchia, Ph.D.
Epidemiology Division
Mario Negri Institute
  of Pharmaceutical Research
Milan, Italy

Ascension Marcos, Ph.D.
Research Associate
Instituto de Nutricion
 y Bromatologia
Facultad de Farmacia
Ciudad Universitaria
Madrid, Spain

Ralph Meer, M.S., R.D.
Department of Nutritional Sciences
University of Arizona
Tucson, Arizona

Jean-Claude Melchior, M.D.
Service de Nutrition
Hopital Bichat — C. Bernard
Paris, France

Richard D. Semba, M.D.
Assistant Professor
Dana Center
Department of Immunology and
 Infectious Diseases
and
Division of Human Nutrition
Department of International Health
Johns Hopkins School
 of Hygiene and Public Health
Baltimore, Maryland

Diego Serraino, M.D.
Epidemiology Unit
Aviano Cancer Centre
Aviano, Italy

Gail Shor-Posner, Ph.D.
Assistant Professor
Department of Epidemiology
 and Public Health
University of Miami
School of Medicine
Miami, Florida

T. Peter Stein, Ph.D.
Department of Surgery
University of Medicine
 and Dentistry of New Jersey
School of Osteopathic Medicine
Stratford, New Jersey

**Cynthia Thomson, M.S., R.D.,
C.N.S.D.**
Clinical Nutrition Research
 Specialist
Department of Family and
 Community Medicine
University of Arizona
Tucson, Arizona

Simin B. Vaghefi, Ph.D.
Department of Nutrition
University of North Florida
Jacksonville, Florida

Pilar Varela, Ph.D.
Research Associate
Instituto de Nutricion y Bromatologia
Facultad de Farmacia
Ciudad Universitaria
Madrid, Spain

Brian J. Ward, M.D.
McGill Center for the Study of
 Host Resistance
Montreal General Hospital
Montreal, Quebec, Canada

Rueben C. Warren, Ph.D.
Centers for Disease Control
Atlanta, Georgia

David R. Woodard, M.D.
Department of Medicine
Section of Infectious Diseases
Eisenhower Medical Center
Rancho Mirage, California

TABLE OF CONTENTS

Nutrition and AIDS:
An Introductory Chapter

Simin B. Vaghefi

and

Eduardo A. Castellon-Vogel

WHAT IS AIDS?

There were an estimated 1.5 million individuals infected with human immunodeficiency virus (HIV) in the U.S.[1,2] In 1992, the CDC reported 160,372 deaths. A 10% increase in infection annually with 60% of cases ending in death is projected.[3,4] In 1990, federal expenditure for AIDS-related medical research was $1.31 billion. Comparing this with $1.01 billion spent in research and 777,000 deaths from heart disease or $1.45 billion and 494,000 deaths from cancer provides insight into the importance of HIV infection and AIDS in this country. The ratio of spending per number of deaths is 10 times more for AIDS than for the other deadly diseases such as diabetes, heart disease, and cancer.

Presently we know that AIDS is not transmitted by food or drink or casual contact. It is not a food-borne disease nor can you treat it by eating certain foods.[5]

AIDS illnesses usually include fever, diarrhea, anorexia, weight loss, and, eventually, opportunistic infections. The gastrointestinal (GI) tract is a major target for AIDS related diseases. Several studies have reported AIDS patients with diarrhea, oral and esophageal candidiasis, dysphagia, and odynophagia. Diarrheal illnesses occur in less than 50% of AIDS patients. These complications contribute to malnutrition, which in turn predisposes the patient to malabsorption and sepsis,[6] furthering the effect of protein-energy malnutrition

0-8493-7842-7/94/$0.00+$.50
© 1994 by CRC Press, Inc.

(PEM). There are a number of adverse effects of PEM as reported by Chandra,[7] the most frequent being the occurrence of opportunistic infections with microorganisms such as *Pneumocystis carinii* and *Candida*. Lymphoid tissues show a significant atrophy. The thymus is small. Histologically, there is a loss of corticomedullary differentiation, there are fewer lymphoid cells, and the Hassall bodies are enlarged, degenerated, and, occasionally, calcified. There is a loss of lymphoid cells around small blood vessels in the spleen. In the lymph node, the thymus-dependent areas show depletion of lymphoid cells. Immunocompetence is impaired.

AIDS produces an unexplained neurological syndrome referred to as AIDS encephalopathy that can result in confusion and dementia, which may lead to the patient not remembering how or when to eat.

THE BIOLOGY OF HIV

The human immunodeficiency virus (HIV) is a member of the class of viruses called retrovirus. This retrovirus contains reverse transcriptase, an enzyme that produces DNA from RNA, a reverse process of normal production of RNA and DNA. Although HIV is an RNA virus, its genome is converted to DNA after cellular infection. Retroviral infections are chronic because of the ability of the virus to produce latent or productive infection. Although other cells can be infected, helper T lymphocytes (CD4+) and macrophages are the major cellular reservoirs for HIV.[8]

Productive infection results in the depletion and death of CD4+ lymphocytes, among other cells. Despite the presence of an active immune response to HIV, the infection persists and gradually depletes the body of helper T lymphocytes and other immunocompetent cells. When the depletion is sufficiently severe, the affected individual becomes vulnerable to a variety of infections and tumors and develops the clinical syndrome of AIDS.

There is a reduction in the number of mature fully differentiated T lymphocytes. There is also a reduction in serum thymic factor activity and an increase in the number of null cells and in deoxynucleotidyl transferase activity in leukocytes, which is a feature of lymphocyte immaturity.[7]

Very little is known about the effects of nutritional stresses on immune systems in animals or humans that are already immunosuppressed by retroviral infection, including HIV-infected patients.

Serum antibody responses are generally intact in PEM, particularly when antigens in adjuvant are administered or in the case of those substances that do not evoke T cell response. At the same time, antibody affinity is decreased, which may provide an explanation for a higher frequency of antigen-antibody complexes found in malnourished patients. As opposed to serum antibody responses, secretory immunoglobulin A antibody levels after immunization with viral vaccines are decreased.

The process of phagocytosis is also affected in PEM. Recent work in humans and animals demonstrated that the production of interleukin-1, interleukin-2, and interferon-gamma is decreased in PEM. Respiratory epithelial cells in subjects suffering from PEM have a higher number of bacteria adhering to one cell compared to well-nourished control subjects.

The investigations conducted in laboratory animals have extended these observations to selected nutrient deficiencies. For example, deficiencies of pyridoxine, folic acid, vitamin A, vitamin C, and vitamin E resulted in impaired cell-mediated immunity and reduced antibody responses. Vitamin B_6 deficiency results in decreased lymphocyte stimulation response to mitogens such as phytohemagglutinin. A moderate increase in β-carotene or vitamin A intake may enhance immune responses. Zinc deficiency, both acquired and inherited, is associated with lymphoid atrophy, decreased delayed cutaneous hypersensitivity responses, and lower thymic hormone activity. Wound healing is impaired. Excess zinc also depresses neutrophil function and lymphocyte responses. Copper-deficient animals show a reduction in the number of antibody-producing cells compared with healthy and pair-fed controls.

The type and amount of lipids in the diet has a significant effect on antibody-producing cells, spleen cell cytotoxicity, the natural history of tumors implanted in rodents, and response to *Listeria*.

Potential immunomodulation by nutritional excesses and deficiencies can act as cofactors in human immunodeficient virus infection. Although AIDS is induced by retrovirus infection, cofactors could accelerate or retard development of immunosuppression. The study of nutritional cofactors in human AIDS patients is vastly complicated by genetic differences, nutritional modifications, complex mixes of interactive cofactors, cost, and necessary safety restrictions for human subjects.

This situation has prompted the scientific community to find animals which respond — similar to humans — when infected with HIV. Such a subject of study was found in mice infested with the murine AIDS (MAIDS). The similarities between the murine AIDS and early stages of human AIDS are striking in many respects. The value of the murine AIDS model resides in the rapid evaluation of immune changes caused by potential cofactors, including nutritional changes.

To be more specific, mice infected with LP-BM5 murine retrovirus-induced AIDS develop hypergammaglobulinemia, lymphadenopathy, T cell functional immunodeficiency, and later neurological signs, including paralysis, as well as opportunistic infections. The virus infects macrophages, B cells, and, to a lesser extent, T cells.[9]

MURINE ACQUIRED IMMUNO DEFICIENCY SYNDROME (MAIDS) USED TO UNDERSTAND HIV WASTING SYNDROME

Experiments on MAIDS indicate that changes induced by dietary omega-3 (n-3) lipids and/or moderate energy restriction on the composition of lymphoid cell membranes can modulate the production or response of lymphokines, including immunosuppressive PGE_2, which appears to delay progression of the retroviral infection.

Omega-3 (n-3) fatty acids and low levels of dietary lipids and/or low zinc levels have been found to be quite effective in inhibiting autoimmune disease in mice.

The level of energy and the source of dietary lipids could effectively modulate proliferative response, production of certain lymphokines, prostaglandins, and fatty acid composition in immune cells and thereby affect the survival rate by delaying the progression of MAIDS disease.[15]

AIDS AND NUTRITIONAL INTERVENTION

It is a known fact that the disease originated in Africa where the syndrome is known as "slim" disease. Experts believe one reason AIDS has such a firm foothold in Africa is that PEM is so widespread in people of all ages and in the region.[10] However, it is known that well-nourished people exposed to the virus also become infected with AIDS.

The nutritional intervention for patients with HIV and AIDS has changed since the recognition of the problem as a syndrome in the early 1980's. An example is the following case study:

In 1987, a 29-year-old male homosexual was admitted at University Hospital in New York. Upon admission, the patient complained of nausea,vomiting, diarrhea, and rectal bleeding and weighed 132 lbs. He was readmitted three times with severe weight loss and several infections. On the last admission, he received peripheral parenteral nutrition (PPN) equal to 1,000 kcal/d, which was increased threefold. Two weeks later, total parenteral nutrition (TPN) was started at 1,000 kcal/d, then increased threefold again. He died 10 days later.

In this report, there was a brief mention of "the restoration of immune defenses by proper nutrient intake" as a "cardinal guideline in treating patients with AIDS secondary to malnutrition." This was more of a call for more aggressive intervention than an observation.

Nutrition intervention can be important in preventing the weight loss and malnutrition seen in HIV infection and AIDS. At present, the role of nutrition has been recognized in the treatment of AIDS. In 1987, the case definition for the disease was revised in order to better describe the conditions associated with HIV infection and to simplify reporting of cases. In the new definition, changes included the addition of HIV wasting syndrome, which is character-ized by emaciation and weight loss.

Nutrition intervention is necessary at all stages of HIV infection. During the asymptomatic period, the goal of nutrition counseling is to promote an adequate, balanced diet for weight maintenance and prevention of vitamin and mineral deficiencies. In the later stages of the disease, nutrition recommenda-tions may involve enteral or parenteral nutrition support.[12]

There are three basic types of supplemental nutrition for AIDS patients. These are total parenteral nutrition (TPN), peripheral parenteral nutrition (PPN), and/or tube feeding (enteral nutrition or TF), which are used in aggressive nutritional treatment of AIDS patients.[13]

The effect of TPN has been shown to vary with the underlying clinical problem. Dietary modifications also may be applied successfully to the nutri-tional care of AIDS patients.[8]

In 1989, The American Dietetic Association published its position paper emphasizing:

Nutrition intervention and education [should] be a component of the total health care provided to individuals infected with human immunodeficiency virus.

This intervention should be implemented at all stages of the disease. The American Dietetic Association also supports nutrition research related to the spectrum of human immunodeficiency virus infection.[12]

It is reported that hospitalized AIDS patients can consume only 70% of estimated basal energy needs and 65% of protein needs, which does not account for the increased needs of hypermetabolism associated with acute infection or any physical activity. Hypermetabolism has been found even in clinically stable patients with AIDS and AIDS-related complex (ARC).

A commonly used indicator of nutritional status is serum albumin. The impact on AIDS of serum albumin concentration has been demonstrated. Patients with normal serum albumin levels generally live longer than patients with low serum albumin levels.

Patients with AIDS have moderate to severe metabolic stress similar to that found in other critically ill patients. This stress, coupled with the anorexia and malabsorption associated with the disease, promotes malnutrition. Nausea, vomiting, and decreased appetite can result from fever, infection, gastrointes-tinal disturbances, the side effects of medications, and emotional issues.[6]

The clinician should consider providing TPN or PPN, especially for patients with massive weight loss (greater than 20%) coupled with diarrhea.[13]

Undernourished individuals have impaired immune responses. The most consistent abnormalities are seen in cell-mediated immunity, complement system, phagocytes, mucosal secretory antibody response, and antibody affinity. These abnormalities predispose the patients to contract more infections. It is now recognized that deficiencies of single nutrients such as zinc, iron, vitamin B_6, vitamin A, copper, and selenium also impair immune responses.

Investigators have found that an early sign of malabsorption is low serum vitamin B_{12} levels. In fact, B_{12} deficiencies have been found in seropositive patients who do not have the active disease.[14] Some drug-nutrient interaction may also take place in patients undergoing chemotherapy.

Taber has pointed out that nausea, vomiting, anorexia, depression, weakness, fatigue, disorientation, neurological complications, fever, dyspnea, and mucosal lesions can affect the ingestion of nutrients. She gives suggestions in her article on how to present and prepare meals for AIDS patients with any of the above mentioned symptoms.[14]

Malnutrition is common in hospitalized patients and can result in immune system dysfunction, thereby increasing the risk of infection. Malnutrition may be a factor in the development of opportunistic diseases and progression of the disease process. It has been reported that 88% of AIDS patients were classified as malnourished on the basis of weight history and serum albumin levels; they had compromised somatic and visceral proteins and notable weight loss. Reports of a 62% incidence of weight loss of 10% or more of usual body weight at 6 to 8 months after diagnosis are common by various investigators in AIDS patients. Clearly, malnutrition, as defined by weight loss and hypoalbuminemia, plays a critical role in the progression of AIDS.[6]

In the U.S., nutritional deficiencies are common among IV drug users and chronic alcoholics. The presence of malnutrition in patients with AIDS may be multifactorial. The primary mechanisms include disorders of food intake, GI disturbances, or alterations in intermediary metabolism, and nutrient malabsorption.[2]

Promising studies employing supplemental nutritional intervention have shown increased weight gain and good food tolerance by patients when elemental or peptide-based diets were given at home.[13]

The Task Force on Nutrition support in AIDS reported that fewer than 20% of the major AIDS treatment centers in the U.S. had standard nutrition protocols for patients with AIDS in 1989. The Task Force has recommended that individuals who test positive for HIV infection receive counseling and nutrition support before they become malnourished.[16]

The most important factor in care of AIDS patients is to provide supportive and non judgmental care to establish a trusting relationship. Individuals with AIDS are generally well informed about their disease and play an active role in making decisions regarding the care they receive. Nutrition treatment plans should incorporate the patient's wishes.[17] It is imperative, for a successful treatment, that dietitians develop and maintain good rapport with HIV-infected individuals.[12]

Protocols for nutrition management of AIDS patients need to be established, and aggressive nutrition support and counseling should be available.[16]

Nutritional management may be conceptualized as a mathematical equation wherein nutritional status is a function of intake, absorption, and metabolism. Disturbances in any of the components of this equation can result in PEM.

Nutritional support can improve nutritional status in selected patients, and repletion of body cell mass may be associated with functional improvement. Early assessment, attention to nutritional requirements, and prompt intervention can minimize wasting and replete body cell mass. The goals of nutritional assessment and intervention are improved nutritional status, prolonged survival, and enhanced quality of life.

Patients, free of intestinal diseases, fed for 2 months with percutaneously placed, endoscopically directed gastrostomy (PEG) tubes to provide 500 kcal/d above estimated requirements gained an average of 3 kg and showed a statistically significant increase in total body potassium (TBK), as well as body fat content, and serum albumin level and other factors. For certain patients, TPN may be the only alternative.[2]

In general, many of the current practices in nutrition care already in place can be applied to caring for persons with HIV infection and AIDS. Nutrition assessment and counseling requires timely follow-up in order to adjust for changes in nutritional status.[12] The experience in Northeast Florida has shown that once the patients are diagnosed HIV positive, they are registered and receive nutrition counseling. But they generally fail to attend the educational programs including nutritional assessment and counseling held every three months. This is either due to becoming too sick to continue in the program or moving away or death of the patient. Outreach programs are more successful in prolonging the life of the patients.

HOME CARE

To control health care costs and decrease hospital stays, more AIDS patients are being treated in the home with intravenous therapies. Dietitians need to become more involved in providing nutrition services in the community and through home care agencies.[17]

The procedure to care for AIDS home care patients treated with TPN therapy at home was published in 1988. This article provides detailed instructions for care and advice in how to solve common TPN problems such as prevention of air embolism.[18]

Recent nutritional knowledge can be used to develop designer formulas that may help reduce the occurrence of opportunistic infections in immunocompromised hosts, such as patients with trauma or burns.

AIDS AND WASTING SYNDROME

Since 1987 HIV seropositivity with wasting (weight loss greater than 10% over 2 months) has been classified as AIDS by the Centers for Disease Control.[2]

PEM is a common occurrence in patients with AIDS. Its incidence may be expected to increase as other disease complications are effectively treated and survival is prolonged. Clinical experience has suggested that PEM per se is an independent source of morbidity and may, on occasion, dominate the clinical picture.

PATHOGENESIS OF PEM IN AIDS

Food intake is inhibited indirectly in patients with malabsorptive diseases or systemic infections due to release of specific factors that inhibit appetite at the central nervous system level. These problems may be exacerbated by economic factors or other impediments to obtain food.[8]

Most common patients' complaint is odynophagia due to esophagitis caused by *Candida albicans,* a fungus. Esophageal ulcers of viral, mycobacterial, and neoplastic varieties also affect food intake.

Anorexia may be a side effect of various medications. Neurologic disease may impair appetite or produce swallowing disorders. There is also evidence that unabsorbed micronutrients in the lower bowel (ileum and colon) are associated with signals that decrease appetite.[2]

It is important that underlying infections promoting wasting be treated promptly. Repletion of body cell mass through effective treatment of disease and by proper nutrition is an important consideration, although repletion is difficult, if not impossible, in patients with untreated serious disease complications.

The importance of this progressive condition has been reinforced by the expansion of the AIDS case definition to "HIV Wasting Syndrome" by the Centers for Disease Control and prevention.

Weight loss, as well as low serum albumin levels, is predictive of an increased risk of morbidity and mortality in hospitalized patients. Studies have established that the frequency of complications from malnutrition increases sharply when serum albumin levels fall below 30 g/L.[13]

AIDS patients typically lose 46% of their potassium by the time of death, and this potassium is lost over the last 9 months of life. The patients also lose 34% of their ideal body weight during the last 4 to 5 months prior to death. In studies reported here, lean body mass (estimated from the potassium pool) is greater than the loss in body weight.

The amount of lean body mass is sufficient in itself to be the cause of death. World War II era starvation studies point out that losses of less than 40% of body weight resulted in death.[19]

Studies of body composition in AIDS patients demonstrate that body cell mass depletion is out of proportion to losses of body weight or fat. The timing of death due to wasting is related to the extent rather than the specific cause. Wasting is not a constant phenomenon. The effect of treating infections that promote wasting was shown in a study of ganciclovir* therapy for cytomegalovirus colitis, in which untreated patients underwent progressive wasting, whereas treated patients repleted body mass. TPN had a variable effect upon body composition, with repletion occurring in patients with eating disorders or malabsorption syndromes and progressive depletion occurring in patients with serious systemic infections. Enteral nutrition also can replete body mass in AIDS patients without severe malabsorption. Pharmacologic stimulation of appetite also may lead to weight gain. Nutritional support can improve nutritional status in selected AIDS patients.

BODY COMPOSITION IN AIDS

PEM develops in most AIDS patients at some point during the course of the disease. Unexplained wasting may be the initial finding indicating an immune deficiency.

Trujillo et al obtained the percent weight change by using the following calculation:

$$\frac{\text{actual weight} - \text{usual weight}}{\text{usual weight}} \times 100$$

The key measurement in these studies is the body cell mass, which is the nonadipose tissue cellular mass or protoplasmic mass. This measurement is not affected by alterations in fluid balance as is the determination of lean body mass or fat-free mass measured by anthropometry or bioimpedance techniques.

Total body protein content has been estimated by measuring total body nitrogen with use of neutron activation techniques. AIDS patients were shown to have total body nitrogen depletion.[8]

A cross-sectional study demonstrated depletion of body cell mass as measured by TBK content out of proportion to a loss of body weight. Shortened survival rate correlated with greater degrees of body cell mass depletion. It has been shown that PEM is not universal in AIDS patients. Kotler theorizes that the loss of nitrogen could be correlated to TBK depletion. Micronutrient deficiencies accompany macronutrient deficit.[8]

* Antiviral agent. Usually administered intravenously.

AIDS AND COUNSELING

There are several resources readily available to use when counseling AIDS patients. There is a videotape from the FDA Office of Public Affairs called "Eating Defensively: Food Safety Advice for Persons with AIDS", which alerts AIDS patients and others with a compromised immune system to avoid handling raw animal foods.[22]

There is a book called *Quality Food and Nutrition Services for AIDS Patients*. The authors' objective in writing this manual was "to provide health care professionals with up-to-date, relevant educational materials needed to deliver quality nutrition and food services to AIDS patients". Further chapters provide the more in-depth material needed to increase the dietitian's under-standing of this complex disease process as it relates to nutritional status. Handout pages for problems frequently encountered by the patient with AIDS (e.g., coping with diarrhea, weight loss, altered taste sensation, and dyspnea) are included. It can be called: "A Complete Education Manual".[23]

As soon as a patient has been diagnosed HIV positive, nutritional counsel-ing should be administered.

In 1989, the Task Force on Nutrition Support in AIDS released the "Guidelines for Nutrition Support in AIDS". Those guidelines will be valuable in assisting in the development of nutritional support plan for patients with AIDS.[16]

The issue of who should counsel and whom should be counseled is complex and controversial. At minimum, AIDS and ARC patients should be made aware of the infectious agents present in raw agricultural commodities. Furthermore, patients should be instructed about proper handling and cooking procedures, as well as general hygienic practices. In many cases such instruc-tions do not go beyond the information that should reach the public at large.[20]

> A viewpoint emerging among many AIDS specialists is changing one's lifestyle may delay the onset and progression of the disease — perhaps for quite a long time. Determining which changes you should encourage patients to make, however, requires a clear understanding of how lifestyle affects the immune system.[10]

AIDS AND FOOD SAFETY

Many infectious diseases associated with HIV infection at any stage are related to lifestyle, and some may be food-borne. Food-borne disease unques-tionably poses a life-threatening risk to AIDS and ARC patients.[20] One must take into consideration that AIDS patients and others with a compromised immune system are 20 times more likely to contract *Salmonella* food poisoning and 200 to 300 times more likely to contract food poisoning from *Listeria* bacteria.[21]

HIV infection is present in practically every level of society, from the richest to the poorest. Concern has been raised about the possibility of food contamination by HIV-infected workers at food processing plants and kitchens in general. Khan has emphatically stated that AIDS is not transmitted by food or drinks.[5]

It would be erroneous to conclude that the food processing industry is not affected by the impact of AIDS and by other causes of immune suppression in our population.

The food processing industry is regulated by the Hazard Analysis and Critical Control Point (HACCP), which assures elimination of microbial infection especially by the "heat kill step". Afterward, the food passes through the next stage before entering "protective packaging". An individual with immune suppression due to infection with HIV or juvenile diabetes, asthmatics, recipients of kidney, heart, lung, and other organ transplants, patients with cancer of a wide variety, and others that might have an opportunistic infection could contaminate the processed food.[5] Such infections as *Salmonella* and *Pneumocystis carinii* are common among immunosuppressed persons.

Kahn recommends the following safeguards against the opportunistic infections while protecting individuals' rights:

1. Safeguard the public by providing wholesome food.
2. Protect the co-workers.
3. Protect the immunosuppressed worker from harm.
4. Assure compliance with the laws that protect these and other handicapped workers.

AIDS AND WOMEN

AIDS poses an increasing threat to infant health worldwide. In an antibody serosurvey of 5,099 people in Zaire, the age groups with the highest rates of infection with HIV were men and women of reproductive age and infants. This age distribution is representative of the World Health Organization's (WHO) Pattern II countries — where the principal routes of transmission are heterosexual and perinatal.[24]

By 1990 there were 1,500,000 reproductive age women already being infected worldwide and over 1,000,000 in Africa alone.[24] The CDC reported in 1992 an estimated 11% of all AIDS cases were females in the U.S.

Women being the care givers to the family may forget their own nutrition. Economics of the family may contribute to the PEM in women. Nutrition education and counseling is important for this group of individuals to combat the disease and prolong their lives.

AIDS AND CHILDREN

A recent study shows that the birth weights and gestational ages of HIV-infected and HIV-seroreverted children were not significantly different. Their weight percentiles diverged significantly by age 19–21 months of age, with preservation of height.[30] Lean body mass, although within the normal range, was significantly different from the HIV-negative, comparison children.

Growth is stable until antiretroviral therapy is required.

Good nutrition is essential in helping children grow and develop. It is especially important for children with HIV.

The fact that nutrient deficiencies of various types are immunosuppressive, especially in children, is well recognized. Good nutrition can also help these children maintain strength, fight infections, and prevent weight loss. Weight loss is one of the most common and challenging health problems facing children with HIV.

A diet containing a variety of foods will help children meet nutritional needs. Feeding a balanced diet to children will reduce the need for a food supplement.[25]

Nutritional support produces benefits beyond simply repleting body mass. It has been associated with marked functional improvement. TPN-induced weight gain has been documented in a pediatric population.[8] In the pediatric population with AIDS, malnutrition may lead to growth failure. It has been demonstrated that, in a subpopulation of severely ill, HIV-seropositive infants with lymphocytic interstitial pneumonitis and failure to thrive, careful monitoring of food intake, frequent feeding, hypercaloric formulas, and nocturnal nasogastric feeds resulted in intake of the recommended amount of calories and adequate rate of weight gain.[2]

AIDS AND BREAST-FEEDING

The CDC estimates an approximate 6,000 births per year by HIV-infested women.[4] HIV has been isolated in breast milk. There are documented cases of virus transmission to infants through breast milk. However, HIV transmission rate via breast-feeding has not yet been determined with precision.

> The American Academy of Pediatrics recommends in the United States and other countries where safe nutrition other than breast-feeding is available, HIV-infected mothers should be advised against breast-feeding their infants to avoid that possible route of HIV infection.

The World Health Organization recommends that HIV-positive women continue to breast-feed wherever a safe alternative is unavailable.[24]

How safe is safe enough? Wet nursing and bottle feeding are classified together as "alternative infant feeding practices". When culturally accepted and feasible in developing countries, wet nursing by a non-HIV infected woman may have a significantly lower risk of infant mortality from non-HIV causes than bottle feeding. There are immunologic and nutritional benefits of breast-feeding, and contamination of bottle formula is avoided. Concerns for transmission of HIV from child to the female wet nurse by breast-feeding have been raised, but such a route of transmission has not been documented.[24]

AIDS AND MEGADOSES

It is speculated that high nutrient intake modulates immune functions and could potentially enhance resistance to progression to AIDS.[9] There are many nutritional therapies to treat AIDS. Some are scientifically sound, while others are not. Despite the amount of money invested in finding a cure for the disease, the drugs developed so far extend the life expectancy but do little for the quality of life. Some AIDS patients, in their desire to survive, use megadoses of vitamins; others take a holistic approach using herbs; and others volunteer to test the latest ideas even if they might be dangerous. Many individuals use supplements to attempt to enhance or prevent loss of immune cells and function, with only limited success.[1]

Low levels of zinc — a mineral needed for immunocompetence — have been reported among apparently healthy male homosexuals.

Because of their desire to gain control of their health, people with AIDS are highly susceptible to fraudulent therapies and remedies. Martin et al, Dwyer, and others have found that one such therapy is megadosing with certain vitamins and minerals.[16,27,26] A megadose level of a nutrient has been defined as more than 10 times the RDA.

Martin et al have found that megadosing with certain nutrients may stimulate or depress immunity and increase susceptibility to infectious challenge. They have also found that the highest consumption levels were for vitamins C and E, zinc, and iron. It must be kept in mind that at certain levels, vitamins and minerals could have a detrimental effect on the human body.

Vitamin C in large doses may cause kidney stone formation and rebound scurvy. Vitamin E is the least toxic of the fat-soluble vitamins; most adults can tolerate oral doses up to 800 mg/d without gross signs or biochemical evidence of toxicity. Excessive consumption of iron and zinc (greater than 20 times the RDA for 6 weeks) impairs immune function and should be avoided.[26]

CAROTENOID EFFECTS

Since carotenoids have received much publicity as anti-oxidants, many AIDS patients have taken them in excessive amounts in desperation. But, high levels of vitamin A as well as its deficiency modulate immune defenses.[9] Carotenoids may also be harmful if taken in excess.

It is practically general knowledge that carotenoids, which are vitamin A precursors, are less toxic and have immunostimulatory activities. Carotenoids have also been implicated as chemoprotective or chemopreventive agents in several kinds of cancer, particularly skin cancer. Some carotenoids such as β-carotene can be converted to vitamin A, which can stimulate the immune system. β-Carotene given to AIDS and oral leukoplakia patients stimulated natural killer cells and activated lymphocytes as it did in normal uninfected volunteers.

β-carotene is a common dietary constituent in many vegetables. It can be metabolized to retinol or vitamin A and is associated with very little toxicity. A study on a trial of β-carotene in subjects infected with HIV[28] has shown that β-carotene is a nontoxic carotenoid with immunomodulating properties in animals and humans. β-Carotene can modulate certain immune markers in HIV-infected subjects.

The effects of β-carotene observed in normal immunocompetent subjects have been an increase in natural killer (NK)[4] cell number and expression of activation markers such as Ia antigen, interleukin-2 receptor, and transferrin receptor. These increases appeared to be maximal after 3 months of β-carotene administration and diminished during the fourth month, even though subjects remained on the drug. Similar findings have been noted in studies of subjects not infected with HIV.[28]

Dwyer et al state:

> Compassionate care does not and should not stop when curative treatment is unavailable...the best job we do at this, the less likely our patients will turn to potentially harmful unproven therapies. Hippocrates said it best: 'I will prescribe regimen for the good of my patients according to my ability and my judgment and never do harm to anyone...'[27]

AIDS AND STRESS

Stress can also compromise immune status. Being under pressure, loneliness, and depression have been shown to cause immunosuppression.

A study by the University of California, San Francisco, Biopsychosocial AIDS Project reported that stressful events frequently precede the diagnosis of AIDS, suggesting that they may have triggered its onset.[10]

REFERENCES

1. Watson, R.R. (1992) Nutrition, immunomodulation and AIDS: an overview, *J. Nutr.* 122: 715.
2. Hecker, L.M. and Kotler, D.P. (1990) Malnutrition in patients with AIDS. *Nutr Rev.* 48: 393–401.
3. Sanlo, R. (1992) *HRS district four AIDS surveillance report,* North East Florida.
4. (1993) Projections of the number of persons diagnosed with AIDS and the number of immunosuppressed HIV-infected persons — United States, 1992–1994. *J Am Med Assoc.* 269: 733.
5. Khan, P. (1990) Food safety in an era of immunosuppression: Is it a problem in the food processing environment? *Nutr Today* 25: 16–20.
6. Ysseldyke, L. (1991) Nutritional complications and incidence of malnutrition among AIDS patients. *J Am Diet Assoc.* 91: 217–218.
7. Chandra, R.K. (1992) Nutrition and immunoregulation. Significance for host resistance to tumors and infectious diseases in humans and rodents. *J. Nutr.* 122: 754–757.
8. Kotler, D.P. (1992) Nutritional effects and support in the patients with acquired AIDS. *J Nutr.* 122: 723–727.
9. Watson, R.R. (1992) LP-BM5, a murine model of acquired immunodeficiency syndrome: role of cocaine, morphine, alcohol and carotenoids in nutritional immunomodulation. *J Nutr.* 122: 744–748.
10. Cerrato, P. (1991) Nutrition and HIV: what to tell your patients. *RN.* 54: 73–78.
11. (1988) Severe malnutrition in a young man with AIDS. *Nutr Rev.* 46: 126–132.
12. (1989) Position of the American Dietetic Assoc.: nutrition intervention in the treatment of HIV. *J Am Diet Assoc.* 89: 839–841.
13. Trujillo, E.B., Borlase, B., Bell, S., Guenther, K., Swails, W., Queen, P., and Trujillo, R. (1992) Assessment of nutritional status, nutrient intake, and nutrition, support in AIDS patients. *J Am Diet Assoc.* 92: 477–478.
14. Taber, J. (1989) Nutrition in HIV infection. *Am J Nursing.* 89: 1446–1451.
15. Fernandes, G., Tomar, V., Venkatraman, M.N., and Venkatraman, J.T. (1992) Potential of diet therapy on murine AIDS. *J Nutr.* 122: 716–722.
16. Task Force on nutrition support in AIDS (1989) Guidelines for nutrition support in AIDS. *Nutr Today.* 24: 27–32.
17. Resler, S.S. (1988) Nutrition care of AIDS patients. *J Am Diet Assoc.* 88: 828–832.
18. Johndrow, P.D. (1988) Making your patient and his family feel at home with T.P.N. *Nursing.* 18: 65–69.
19. (1990) What do we know about the mechanisms of weight loss in AIDS?. *Nutr Rev.* 48: 153–155.
20. Archer, D.L. (1989) Food counseling for persons infected with HIV: strategy for defensive living. *Pub Health Rep.* 104: 196–198.
21. Campbell, S. (1991) Eating defensively: food safety advice for persons with AIDS (video review). *Nutrition Today.* 26: 40.
22. Shealy, C.N. (1991) Kuperman, S. (1989) Eating defensively: food safety advice for persons with AIDS; article review. *J Nutr Ed.* 26: 40.

23. Reddick, J. (1991) Quality food and nutrition services for AIDS patients (book review). *J Nutr Ed.* 23: 193–194.
24. Hyemann, S.J. (1990) Modeling the impact of breast-feeding by HIV-infected women on child survival. *Am J Pub Health.* 80: 1305–1309.
25. (1988) *HIV, nutrition, and your child*, U.S. Department of Health and Human Services.
26. Martin, J.B., Easley-Shaw, T., and Collins, C. (1991) Use of selected vitamin and mineral supplements among individuals infected with HIV, *J Am Diet Assoc.* 91: 476–478.
27. Dwyer, J.T., Bye, R.L., Holt, B.S., and Lauze, S.R. (1988) Unproven nutrition therapies for AIDS: what is the evidence? *Nutrition Today.* 23: 25–33.
28. Garewal, H.S., Ampel, N.M., Watson, R.R., Prabhala, R.H., and Dols, C.L. (1992) A preliminary trial of beta-carotene in subjects infected with the HIV. *J Nutr.* 122: 728–732.
29. (1989) *AIDS forecasting: undercount of cases and lack of key data weaken existing estimates.* United States General Accounting Office.
30. Miller, T., Evans, S., Orav, E.J., Morris, V., and McIntosh, K. (1993) Growth and body composition in children infected with the human immunodeficiency virus-1. *Am J Clin Nutr.* 57: 588–592.

FURTHER READINGS

1. (1989) The 1980s: a look at a decade of growth in dietetics through the pages of the Journal. *J Am Diet Assoc.* 89: 1742.
2. (1989) Malnutrition and weight loss in patients with AIDS. *Nutr Rev.* 47: 354–355.
3. (1989) Guidelines for nutrition support in AIDS. *J School Health.* 59: 170.
4. (1989) Task force sets nutrition guidelines on HIV infection. *Am Fam Pract.* 39: 376.
5. (1990) Nutrition handbook for AIDS (book review). *J Nutr Ed.* 22: 253.
6. (1990) Healthy people 2000. *N Engl J Med.* 323: 1065.
7. (1991) Treating malnutrition in AIDS: comparison of dietitians' practices and nutrition care guidelines. *J Am Diet Assoc.* 91: 1273.
8. (1991) Nutrition and HIV (editorial). *Lancet.* 338: 86.
9. (1991) Breast-feeding in 1991 (editorial). *N Engl J Med.* 325: 1036.
10. (1991) Helping A.I.D.S. patients eat. *Nursing.* 21: 74.
11. (1992) Providers develop specialized AIDS home care programs. *Hospitals.* 66: 144.
12. (1992) Acquired immunodeficiency syndrome — 1991. *J Am Med Assoc.* 268: 713.
13. Bandy, CH.E. (1993) Nutrition attitudes and practices of individuals who are infected with HIV and who live in South Florida. *J Am Diet Assoc.* 93: 70.
14. Miyamura, J.B. (1991) Quality food and nutrition services for AIDS patients (book review). [Sherman, C., Raucher, B., Epstein, J., and Berger, M., Aspen Publishers, Frederick, MD] *J Am Diet Assoc.* 91: 1349.
15. Watzl, B. and Watson, R.R. (1992) Role of alcohol abuse in nutritional immunosuppression, *J Nutr.* 122: 733–737.

Chapter

2

Nutrition and AIDS in Africa

Cyril O. Enwonwu

and

Rueben C. Warren

INTRODUCTION

In every population so far studied, the acquired immunodeficiency syndrome (AIDS) is preceded with infection by a retrovirus known as human immunodeficiency virus type 1 (HIV-1) which triggers an irreversible immune dysfunction. Replication of HIV-1 starts after the virion enters a host cell, and once internalized, the viral reverse transcriptase synthesizes DNA from the viral RNA. The double-stranded viral DNA becomes integrated into the host's genome.

Like many diseases, the natural history of HIV infection in Africa seems to differ from what is observed in the affluent, technically developed world. In Western Europe and North America, 50% of HIV-infected individuals develop AIDS within 10 years.[1] For reasons not completely understood, the period of clinical latency separating the early asymptomatic stage of HIV infection from the later progressively symptomatic stages is relatively brief in Africa, with the infection running a fulminant downhill course.[2,3] Factors suspected to influence the duration of clinical latency include genetic susceptibility to HIV infection, the virus load, concurrent infections with other immuno-suppressive agents, and the pre-existing immune status at time of HIV infection.[1,4,5] To enhance our comprehension of the disease process and thus promote the design of appropriate management modalities, the roles of potential codeterminants of the progression from HIV infection to AIDS need to be unraveled.

0-8493-7842-7/94/$0.00+$.50
© 1994 by CRC Press, Inc.

On a global basis, malnutrition is the most widespread immunosuppressive in the human. Malnutrition in Africa is of staggering magnitude, and getting worse in some countries, particularly in sub-Saharan Africa.[6,7] The dominant pathologies in Africa are immunosuppressive transmissible diseases of parasitic and infectious nature (malaria, measles, ascariasis, schistosomiasis, tuberculosis, filariasis, the diarrheas,respiratory infections and others), which are all compounded by the common denominator of malnutrition.[8-10] There is marked overlap between the immunological abnormalities caused by malnutrition and by HIV infection.[11-13] Therefore, an underlying assumption of this report is that pre-existing malnutrition is disadvantageous to the HIV-infected individual and influences the biological gradient and natural history of the viral infection in Africa. An earlier review had examined some aspects of the interface of malnutrition and HIV infection in sub-Saharan Africa.[14]

GEOGRAPHY AND DEMOGRAPHY OF AFRICA

Africa, the second largest continent in the world, occupies about 25% of the world's land mass and contains no less than 52 politically independent countries. These countries constitute a world of bewildering contradictions. Africa is a relatively insecure and politically unstable continent of profound paradoxes in which the majority residing in the rural areas and poor urban towns lack the basic essentials of life, while the very few rich, many of whom are often educated in the best traditions of the Western World, play golf and polo with style and simulated Englishness.

Although statistical data are scarce and largely crude, the total population of Africa was estimated to be more than 500 million in 1984.[15] Sub-Saharan Africa alone, with a population of more than 400 million in 1985,[16] contains 45 countries. The projected population of sub-Saharan Africa by the end of the present century is 678 million. No less than 50–60% of the populations of African countries are under 19 years of age. Annual birth rate varies between 2.5–4.5% in the different countries, with infant mortality rates (120–160 per 1,000) 10- to 20-fold higher than in many developed countries. Children less than 5 years of age constitute 20% of the total population in sub-Saharan Africa but account for 60% of all deaths. Life expectancy at birth varies from 50–56 years for both males and females.

The majority of Africans (65–85%) still reside in rural communities. In 1965, the urban population of Africa stood at 48 million, rising to 103 million in 1980. With the current estimated mean urban growth rate of 5–10%, it is projected that about half of the total population of Africa will be city/urban dwellers by the year 2000. The present tilt of population balance in favor of the urban areas in the developing world, particularly in Africa, has been appropriately characterized by the United Nations demographers as "the greatest mass migration in human history".[17] The rapid urbanization has serious implications with respect to the spread of HIV infection in the continent.[18-20]

Virtually all the black African countries face serious developmental problems. These countries are characterized by economic poverty, rapidly deteriorating plight of agriculture, widespread chronic malnutrition and undernutrition, escalating external debt crisis, grossly inadequate safe water supply,[21] mounting pressures of rapid population growth rate, and very low per capita gross national product. In African countries, per capita annual expenditure on public health is about $4.00 (U.S.) compared to $1,200 in the U.S.

CURRENT STATUS OF HIV INFECTION IN AFRICA

HIV-1 and HIV-2 are the two strains of retroviruses identified in Africa, with the less pathogenic HIV-2 more common in West Africa than in East, Central, and Southern Africa. Body fluids, specifically blood and semen, remain the major sources of HIV. The mode of transmission of HIV in Africa is mainly through heterosexual intercourse, vertical congenital fetal perinatal infection and blood transfusion.[1,20,22] Information about the AIDS epidemic in Africa is still largely crude. Although just over 9% of the world's population reside in sub-Saharan Africa, the region is believed to harbour 60–80% of all HIV-infected individuals worldwide.[23] The World Health Organization estimates that no less than 6 million (including about 500,000 infants and children) of the 8 to 10 million persons infected with HIV worldwide during the first decade of the AIDS epidemic resided in sub-Saharan Africa, excluding South Africa.[1,19] With the start of the second decade, HIV infection in the continent is distributed extensively within the general adult population, particularly in East, Central and Southern Africa, and the concept of isolated high-risk groups as applied to the developed countries is not meaningful.[3]

The present estimated prevalence of HIV infection in women aged 15–49 years in sub-Saharan Africa is 2,500 per 100,000,[3,24] and between 20–40% of these women will transmit the virus to their offspring.[20] By 1991, approximately 180,000 AIDS cases were children aged 0–4 years in sub-Saharan Africa.[25] The WHO predicts 14 million cases of HIV infection in Africa by the year 2000. Children who traditionally constitute the group most vulnerable to malnutrition and to the various endemic tropical infections in Africa will bear an increasingly disproportionate share of the continent's HIV/AIDS burden during the second decade of the epidemic.[3,18,24]

UNIQUE FEATURES OF AIDS IN AFRICA

Striking differences are observed between the clinical manifestations of HIV infection/AIDS in Africa and in the relatively affluent western countries, and even between various population groups in the continent.[26] This may relate not only to genetic differences but also to the varying background of such ecological variables as endemic infections and the status of nutrition.

There are marked racial differences in the ability to handle exposure to infections and other environmental insults.[27,28] Neutrophils, which are avidly phagocytic cells, constitute the first line of defense against microbial invasion of the tissues. Examination of race as a variable in the parameters of immunologic and inflammatory status of healthy individuals with early-onset infectious diseases reveals a significant difference in neutrophil (PMN) chemotaxis between whites and blacks, with the former group showing a higher response.[28] Similarly, antibody responses to some bacteria are influenced by race.[29] It is not yet clear whether such differences are due to genetic or environmental factors such as the status of nutrition.

Common alleles (variants at a single locus) or polymorphisms constitute the basis of human diversity, including the ability to cope with hostile environmental challenge.[30] There is some evidence that genetic variables influence susceptibility to HIV infection and the subsequent progression to AIDS. In Central African countries where HIV-1 infection is very prevalent, the group specific component Gc/fast (Gc/f allele) predominates in the indigenous populations.[31] Gc, an α-2 globulin found both in serum and on cells, is a vitamin D binding factor, and when bound to this vitamin, can transport large amounts of calcium. Studies demonstrate that progression from HIV infection to AIDS has a strong positive correlation with the Gc/f allele.[30]

A number of ecological and behavioral variables intensify the risk of gastrointestinal diseases in most parts of Africa.[8] These include extremes of atmospheric humidity exceeding 90% during the rainy season, high mean annual temperatures, reliance on grossly inadequate heavily contaminated water supply, widespread lack of refrigeration and other modes of food preservation, close proximity of livestock, earth-floored residential units in many impoverished communities, and very poor disposal of human and animal fecal materials.[32,33] It is not, therefore, surprising that gastrointestinal symptoms, particularly diarrhea, occur in 30–50% of North American and European patients with AIDS, and in nearly 90% of AIDS patients in Africa.[10,34] "Slim Disease", the devastating symptom complex of severe diarrhea and cachexia often encountered in Africans with AIDS, may in part be due to synergism between intestinal infections elicited by opportunistic pathogens and pre-existing underlying gastrointestinal infections caused by traditional tropical pathogens.[10,34] Oral as well as esophageal candidiasis, Kaposi Sarcoma lesion usually occurring on the hard palate, severe periodontal diseases, bilateral enlargement of the parotid gland, xerostomia, and mycobacterial infections occur frequently in AIDS patients in Africa while Pneumocystis carinii pneumonia is reported to be less commonly found than in patients in the developed temperate zones.[10,26] Leonidas[35] has observed that in blacks with AIDS, the hair is longer, lighter, softer and occasionally discolored. The AIDS trichopathy is reminiscent of hair changes frequently encountered in protein energy-deficient African children, and attributed to free radical damage among other possible causes.[36]

AIDS is essentially a clinical disorder representing the end point in a progressive sequence of immunosuppressive impairments that render the host very susceptible to life-threatening tumors and opportunistic infections.[37,38] The rapid course and clinical expression of HIV infection in African patients may therefore be due to antecedent or concurrent infection with other endemic immunosuppressive viruses (e.g. Epstein-Barr Virus, Cytomegalovirus, Measles)[4] and parasitic diseases (e.g. malaria, schistosomiasis).[39,40]

DIET AND NUTRITION IN AFRICA: CURRENT STATUS

Certain foods are preferentially utilized in specific African countries and ethnic communities. Nonetheless, there are striking similarities in the staple foods and dietary habits of Africans regardless of their country of origin and their socioeconomic status. The important staple foods eaten by Africans and the protein-calorie ratios (percent) of these staples are wheat (11%), millet (12%), rice (8%), maize (10%), taro (7%), yam (6%), plantain (4%) and cassava (3%). The major meals are bulky, and the amount of staples usually eaten in a day do not provide the RDA of energy, protein and other essential nutrients.[41] Africa has an extensive range of protein-rich foods (e.g. beans of various kinds, cowpeas, groundnuts, sesame), foods rich in vitamins and minerals (green leafy vegetables, banana, mango), and energy-rich foods (palm oils, groundnut oil), but most of these are either expensive or underestimated by the people.[41,42] Plain cereals and cassava porridge feature prominently as foods for children and the sick.[41]

Breast-feeding, both as a source of high quality, easily digestible nutrients and as a means of protection from common infections, is critical for child survival in developing countries, and is widely practiced during the first two years of life in most African communities.[41,43] There is, however, good evidence of severe decline in incidence and duration of breast-feeding among the urban populations in recent years.[43] After the age of 3–6 months, breast-feeding alone does not support normal growth and therefore complementary foods should be introduced. The traditional complementary foods used in Africa are generally of low energy density, low protein and micronutrient contents, and are frequently contaminated with high loads of microorganisms.[41] These complementary foods are often prepared and served under poor hygienic conditions, thus setting the stage for repeated episodes of diarrhea. Anorexia and gastrointestinal malabsorption are frequently associated with the diarrhea, thus aggravating the poor nutritional status.

Because of the rare isolation of HIV from human milk, and anecdotal reports of apparent postnatal transmission of the virus from mothers to their breast-fed children, the role of breast-feeding by HIV-positive mothers has become controversial.[44,45] Studies suggest that the estimated risk of HIV-1 transmission through breast-feeding is 29% if a mother acquires the virus after the child is born, but if the mother is infected prenatally, the additional risk of transmission in utero or during delivery is 14%.[46] The WHO/UNICEF expert consultation[47] and other investigators[46,48] have concluded that in situations similar to those that prevail in Africa where infectious diseases and malnutrition are the major causes of the high infant mortality rate, breast-feeding should be the recommendation regardless of the HIV status of the mothers. This view is supported by recent observations of the presence in milk and colostral samples of a factor which inhibits binding of the HIV epitope-specific MAB to recombinant CD4 receptor molecules as well as the binding of gp 120 to CD4, with no difference in titers of inhibitor activity between samples from HIV-seropositive and -seronegative mothers.[45] A similar inhibitory activity has been reported in human saliva.[49,50]

For most African countries, the first decade of the AIDS epidemic was a period of severe economic crisis characterized by unsustainable foreign debt burden, monetary devaluation, and a ten- to twenty-fold increase in consumer price indices with no real increase in wages.[51-53] Annual population growth rate was greater than that of food production.[54,55] In Zambia, for example, with urbanization in the country close to 60%, most people depend on wages to buy foods. The consumer price index in Zambia rose from 125 in 1976 to 1,625 in 1988, while real wages showed a slight decrease.[7] Ethiopia alone had no less than 9 million famine victims in 1983, and that ominous situation is today being recreated in Somalia and other African countries.

Common nutritional problems in Africa include protein-energy malnutrition (PEM) (usually complicated by concurrent deficiencies of several micronutrients), the nutritional anemias and deficiencies of iodine, zinc, vitamins A and B_2 among others.[56] In 1985 alone, an estimated 25–30 million African children were affected by PEM,[7] and the situation is getting worse.[6]

MALNUTRITION IN HIV INFECTION

The epidemiology of malnutrition and of infectious diseases is inextricably intermingled particularly in impoverished communities in the Third World.[10,11,56] Poor diet and malnutrition impair several parameters of the host's specific and nonspecific defense systems resulting in increased susceptibility to infections, and the latter, in turn, intensify the state of malnutrition (Figure 1). The extensive interaction between nutritional status, immune function, and infections is the subject of several excellent reviews,[11,12,37,57] and will not be discussed in this report.

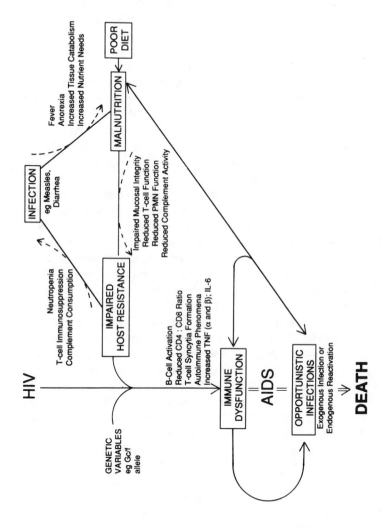

Figure 1. Schematic representation of the complex interactions between malnutrition, host resistance and the common tropical pathogens, and how these in concert with genetic factors impact the transition from asymptomatic HIV infection to AIDS. (From Enwonwu, C.O., *Nutr. Res.*, 12, 1041, 1992. With permission.)

Several reports have underscored the observation that the clinical features of HIV-infected children in Africa are very similar to those observed in general tropical pediatrics.[26,58] This militates against the sole use of clinical criteria to separate severe malnutrition from pediatric AIDS in Africa. Schuerman and colleagues[58] have noted that the severity of malnutrition and other clinical criteria (e.g. failure to thrive, chronic diarrhea, generalized lymphadenopathy and oropharyngeal candidiasis) were comparable between 30 HIV-positive and 64 HIV-negative children in rural Equatorial Africa. There is as yet no convincing evidence that a pre-existing state of malnutrition promotes the acquisition of HIV infection. Since normal immune function is dependent on good nutritional status,[11,57] several investigators have advanced the hypothesis that malnutrition is the predominant underlying cause for the full clinical expression of AIDS in HIV-seropositive individuals.[12,13,37,59,60] According to Raiten,[37] inadequate nutrition may influence specific systems involved in the progression from asymptomatic HIV infection to the full-blown condition of AIDS as well as intensify the susceptibility to opportunistic infections and contribute to the severity of HIV-related diseases.

Both hypometabolism[61] and hypermetabolism[62,63] have been reported, depending on the duration of HIV seropositivity and the presence or absence of secondary infections. Tetrapolar body impedance analysis indicates that the malnutrition occurs early in otherwise symptomless HIV-infected individuals (Walter-Reed, WR, Class 2) and it is speculated that this may be due to either increased metabolic requirements or to impaired cellular utilization of nutrients.[64] PEM is a common complication of HIV infection[14,62] and occurs concurrently with deficiencies of several micronutrients.[65-67] Despite generally adequate and often high levels of dietary intake of many essential micronutrients, studies in asymptomatic HIV-1 positive individuals indicate high prevalence of deficiencies of zinc, copper and several vitamins including vitamins A, E, B_2, B_6, B_{12}, C and folate.[66-68] Many of these nutrients function either as antioxidants or as key components of antioxidant enzymes.[36] GSH (γ-glutamyl-cysteinyl-glycine) concentration in plasma is reduced to about 30% of control level in asymptomatic HIV-positive individuals.[69] GSH accounts for over 90% of the intracellular nonprotein thiols and functions as a major cellular antioxidant and an activator of T cells.[70] It is also required for the synthesis of leukotrienes which are important mediators of inflammation. The level of the precursor amino acid, cysteine, is also markedly decreased in the plasma and mononuclear cells of patients with AIDS.[71] The importance of cysteine has been further enhanced by recent reports that this amino acid can function in its own right as a cytokine that regulates functional activity of lymphocytes.[72] It is perhaps relevant to point out that blood GSH level is also significantly reduced to almost 50% of control level in protein energy-deficient (kwashiorkor type) African children who are HIV-1 seronegative.[73] Meat sources and certain vegetables are rich in GSH as well as cysteine, but most foods are, however, low in the amino acid.[65,74] The possible causes of GSH depletion in malnutrition and in HIV infection include dietary insufficiency, increased oxidative stress with formation of GSSG, extensive formation of mixed disulfides, and the drain on the body's pool of sulfur amino acids as a consequence of chronic diarrhea.

GSH reductase is a flavin enzyme and its level is markedly influenced by riboflavin status which is depleted in HIV infection.

Factors implicated in the causation of malnutrition in HIV-infected individuals include anorexia, endocrine imbalance, hypermetabolism (due to tumor cachexia and infection) and use of medications, some of which antagonize the normal metabolism of nutrients.[37,62,75] Diarrhea and intestinal malabsorption are perhaps the two major causes of malnutrition in HIV-infected individuals,[37,62] and the most difficult to treat.[34,76]

RELEVANCE OF NUTRITIONAL FACTORS TO PROGRESSION OF HIV INFECTION

The host's response to infections and inflammatory stimuli is mediated by cytokines, a diverse range of polypeptides that have multiple biological activities including enhancing the recruitment, proliferation, activation, and differentiation of white blood cells, as well as mediating a wide variety of metabolic alterations. Cytokine production and actions are affected by malnutrition.[77] Among the best studied cytokines are interferons (INF) α, β and γ, interleukins-1 to 6 (IL-1 to 6), and the tumor necrosis factors α and β (TNF α and β). INF-α, β and γ, and IL-2 to 4 stimulate and modify immune function. Others like IL-1 and -6, and TNF-α and β, not only influence the immune system but also initiate prominent changes in energy, protein and mineral metabolism.[77,78]

Abnormalities in cytokine production characterize HIV infection, with impaired production of some like INF-γ,[79] and overproduction of TNF and IL-6.[75,80-82] Mean serum level of TNF-α progressively increases across the clinical spectrum of HIV infection with the highest levels noted in patients with marked weight loss and opportunistic infections. Whether these changes are due to the infection per se[81] or to the underlying state of malnutrition[77,83-86] are not clear. IFN-α, β and γ induce host-cell resistance to a wide range of viral infections, including many animal retroviruses, and are the only cytokines widely believed to retard/inhibit HIV replication. Underproduction of IFN should therefore promote the progression of HIV infection. Both TNF[80,81] and IL-6[82] are elevated in HIV-infected individuals, and, indeed, AIDS is believed to be a TNF disease by some investigators[81] since TNF-α and -β are considered crucial in the transcriptional enhancement of HIV replication.[81,87] Similarly, IL-6 is reported to activate HIV post-transcriptionally.[88]

The stimulation of HIV replication by TNF is linked to induction of NF-kB (nuclear factor kB), a DNA binding protein with binding sites in the viral enhancer, originally thought to be a β cell-specific transcription factor, but now shown to be induced in most cell types.[81,89] It is a heterodimeric nuclear protein consisting of 50- and 65-kDa subunits with both subunits participating in DNA binding. In nonstimulated cells, the majority of NF-kB activity is present in the cytosolic fraction as a cryptic form unable to bind DNA until the p:50 p:65

heterodimer is dissociated from a 37-kDa inhibitory polypeptide, IKB.[89] A protease encoded by the HIV-1 infected cells can contribute to activation of NF-kB by processing the precursor p105 to p50.[90] Similarly, excessive cellular production of reactive free radicals, a consistent feature of malnutrition and infections,[36,75,80] can activate NF-kB.[65] The important thiol GSH, whose cellular levels are profoundly depleted in HIV infection[6] and in PEM,[73] as well as N-acetylcysteine, has now been shown to block the ability of TNF and other mitogens to stimulate HIV-1 replication by inhibiting activation of NF-kB.[91,92] Impaired cellular antioxidant status is a consistent prominent feature of PEM[36] and other forms of malnutrition,[83] and may very well play a key role in the rapid replication of HIV in the malnourished.[14,65] Ascorbate and vitamin A are other antioxidant micronutrients whose tissue levels are depleted in PEM[93,94] and in HIV infection.[83,95] It has been demonstrated that in a chronically HIV-infected T lymphocyte cell line, nontoxic concentrations of ascorbate in the cell culture medium reduces the level of extracellular reverse transcriptase activity by 99%, and the expression of p24 antigen by 90%.[96] Additionally, ascorbic acid supplementation causes a time- and dose-dependent reduction in HIV activity in acutely infected CD4+ lymphocytes.[96] Ascorbate spares GSH, and can serve as an essential antioxidant in the presence of severe GSH deficiency.[97]

CONCLUSION

Normal immune function is critically dependent on good nutritional status, and it has been speculated that malnourished Africans, like immunosuppressed homosexual males, drug addicts and hemophiliacs, may constitute a special risk group for AIDS. There is, however, no convincing evidence yet that malnutrition promotes the acquisition of HIV infection. Nutritional status, like genetics, is a major determinant of host resistance to infections, and is now known to play an important underlying role in the full clinical expression of AIDS in HIV-seropositive individuals.[59,60] Factors most likely to promote the rapid transition from asymptomatic HIV infection to AIDS characteristic of many African communities include malnutrition-induced antecedent impairment of the host's immune status (e.g. pre-existing low CD4+: CD8+ ratio, abnormalities in the production and actions of cytokines, and depleted levels of key antioxidant nutrients/enzymes), and concurrent infection with other potentially immunosuppressive viruses and parasites. In effect, the clinical expression of HIV infection is worse when the infection co-exists or is preceded by malnutrition.

The mucosal tissues of the oral cavity and the rest of the gastrointestinal tract have relatively high cell turnover rates. In PEM, a fundamental problem is poor cell production resulting in intestinal mucosal atrophy. In view of the high prevalence of endemic intestinal parasitic infestations in Africa,[8,10] PEM is associated with prominent destruction of villus cells and hyperplasia of the crypt, resulting in reduced villus/crypt ratio.[98] Thus PEM in Africa not only promotes intestinal infections[10,57] and diarrheas but also impairs absorption of several essential nutrients. It is against such pre-existing conditions that one must consider the consequences of superimposed HIV infection which by itself has a major impact on the structure and function of the entire gastrointestinal tract.[75] Similarly, in the oral cavity, malnutrition-induced mucosal disruptions in combination with poor oral hygiene observed in impoverished Africans[33] are expressed as angular cheilitis, stomatitis, candidiasis and severe periodontal lesions.[33,99] There is also hypofunction of the salivary glands in PEM resulting in xerostomia and therefore inadequate provision of salivary factors required to protect the oral tissues against the numerous potentially pathogenic oral microbial organisms.[100,101] It is not, therefore, surprising that oral lesions such as candidiasis and periodontal diseases are generally more severe in HIV-infected Africans than in similar patients in Western Europe and North America.

An aggressive nutrition program is mandatory for the management of HIV infection. Guidelines for this have been reported in earlier publications.[14,102,103] Implementation of these guidelines poses a formidable task in impoverished Africans infected with HIV in view of the pre-existing state of malnutrition and the complications of other endemic diseases peculiar to the tropical world.

REFERENCES

1. Blattner, W.A. HIV epidemiology: past, present and future. *FASEB J.* 1991; 5:2340–48.
2. Anzala, A., Wambugu, P., Bosire, M., Ngugi, E.N., Waiyaki, P., and Plummer, F.A. The rate of development of HIV-1 related illness in women with a known duration of infection. 1990; International Conference, AIDS 6, 143 (Abstr. No. Th. C. 37).
3. Berkley, S. HIV in Africa: what is the future. *Ann. Intern. Med.* 1991; 116:339–341.
4. Tindall, B. and Cooper, D.A. Primary HIV infection: host responses and intervention strategies. *AIDS* 1991; 5:1–14.
5. Nelson, J.A., Ghazal, P., and Wiley, C.A. Role of opportunistic viral infections in AIDS. *AIDS* 1990; 4:1–10.
6. Berg, A. Sliding toward nutrition malpractice: time to reconsider and redeploy. *Am. J. Clin. Nutr.* 1992; 57:3–7.
7. Maletnlema, T.N. Politics and nutrition in Africa. *World Health* 1991; July/August: 14–15.
8. Pawlowski, Z.S. Implications of parasite-nutrition interactions from a world perspective. *Fed. Proc.* 1984; 43:256–260.
9. Gentilini, M. Co-operation between North and South. (in Wood, C. and Rue, Y., eds.) *International Congress and Symposium Series,* Number 24, pp 1–6, 1980; Academic Press Inc. (London) Ltd. and The Royal Society of Medicine.
10. Cook, G.C. Tropical medicine. *Postgrad Med. J.* 1991; 67:798–822.

11. Chandra, R.K. Nutrition and immunity: lessons from the past and new insights into the future. *Am. J. Clin. Nutr.* 1991; 53:1087–1101.

12. Jain, V.K. and Chandra, R.K. Does nutritional deficiency predispose to acquired immune deficiency syndrome? *Nutr. Res.* 1984; 4:537–543.

13. Sidhu, GS and El-Sadr, W. Some thoughts on the acquired immunodeficiency syndrome (AIDS). *Nutr. Res.* 1985; 5:3–7.

14. Enwonwu, C.O. Interface of malnutrition and human immunodeficiency virus infection in sub-Saharan Africa: a critical review. *Nutr. Res.* 1992; 12:1041–1050.

15. Barmes, D. and Tala, H. Oral health trends in WHO African region. *Afr. Dent. J.* 1987; 1:2–4.

16. McNamara, R.S. The challenges for sub-Saharan Africa. Sir John Crawford Memorial Lecture, Washington, D.C., Nov. 1, 1985.

17. Newland, K. *The city limits: emerging constraints on urban growth.* Washington, D.C.: Worldwatch Institute, 1980.

18. Ulin, P.R. African women and AIDS: negotiating behavioral change. *Soc. Sci. Med.* 1992; 34:63–73.

19. Palca, J. The sobering geography of AIDS. *Science* 1991; 252:372–73.

20. Chin, J. Current and future dimensions of the HIV/AIDS pandemic in women and children. *Lancet* 1990; 336:221–24.

21. Isely, R.B. Water supply and sanitation in Africa. *World Health Forum* 1985; 6:213–19.

22. Lucas, S.B. AIDS in Africa. Clinicopathological aspects. *Trans. R. Soc. Trop. Med. Hyg.* 1988; 82:801–802.

23. Caldwell, J.C., Orubuloye, I.O., and Caldwell, P. Underreaction to AIDS in sub-Saharan Africa. *Soc. Sci. Med.* 1992; 34:1169–82.

24. Quinn, T.C., Ruff, A., and Halsey, N. Pediatric acquired immunodeficiency syndrome: special considerations for developing nations. *Pediatr. Infect. Dis. J.* 1992; 11:558–568.

25. Chin, J., Sato, P.A., and Mann, J.M. Projections of HIV infections and AIDS cases to the year 2000. *Bull. WHO* 1990; 68:1–11.

26. Harries, A. The clinical spectrum of HIV infection in Africa. *Africa Health* 1989; 11:35–39.

27. Scott, J. Molecular genetics of common diseases. *Br. Med. J.* 1987; 295:769–71.

28. Schenkein, H.A., Best, A.M., and Gunselley, J.C. Influence of race and periodontal clinical status on neutrophil chemotactic responses. *J. Periodont. Res.* 1991; 26:272–275.

29. Gunsolley, J.C., Twe, J.G., Gooss, C.M., Burmeister, J.A., and Schenkein, H.A. Effects of race and periodontal status on antibody reactive with Actinobacillus actinomycctcmcomitans strain Y4. *J. Periodont. Res.* 1988; 23:308–312.

30. Eales, L.J., Parkin, J.M., Forster, S.M., Nye, K.E., Weber, J.N., Harris, J.R.W., and Pinching, A.J. Association of different allelic forms of group specific component with susceptibility to and clinical manifestation of human immunodeficiency virus infection. *Lancet* 1987; 11:999–1002.

31. Constans, J., Viau, M., Cleve, H., Jaeger, G., Quilici, J.C., and Palisson, M.J. Analysis of the Gc polymorphism in human populations by isoelectric focusing

on polyacrylamide gels. Demonstration of sub types of the Gcl allele and of additional Gc variants. *Human Genet.* 1978; 41:53–60.

32. Etkin, N.L. and Ross, P.J. Food as medicine and medicine as food. *Soc. Sci. Med.* 1982; 16:1559–73.

33. Enwonwu, C.O. Infectious oral necrosis (cancrum oris) in Nigerian children: a review. *Community Dent. Oral Epidemiol.* 1985; 13:190–194.

34. Smith, P.D. Gastrointestinal infections in AIDS. *Ann. Intern. Med.* 1992; 116:63–77.

35. Leonidas, J.R. Hair alteration in Black patients with acquired immunodeficiency syndrome. *Cutis* 1987; 39:537–38.

36. Golden, M.H.N. The consequences of protein deficiency in man and its relationship to the features of kwashiorkor. In: Blaxter, K. and Waterlow, J.C., eds. *Nutritional Adaptation in Man,* London: John Libbey, 1985:169–85.

37. Raiten, D.J. Nutrition and HIV infection. FDA Contract No. 223-88-2124, Task Order No. 7, Bethesda, MD: FASEB Life Sciences Research Office, 1990; 3–47.

38. Welsby, P.D. Infection and infectious diseases. *Postgrad. Med. J.* 1990; 66:807–817.

39. McGregor, I.A. Malaria: nutritional implications. *Res. Infect. Dis.* 1982; 4:798–804.

40. Duggan, M.B., Alwar, J., and Milner, R.D.G. The nutritional cost of measles in Africa. *Arch. Dis. Child.* 1986; 61:61–66.

41. Hiel, A.M.M., Hautvast, J.G.A.J., and Den Hartog, A.P. *Feeding Young Children.* Nederlands Instituut Voor De Voeding, The Netherlands Nutrition Foundation, Wageningen, 1982.

42. McDowell, J. In defence of African foods and food practices. *Trop. Doctor* 1976; 6:37–42.

43. Anyanwu, R.C. and Enwonwu, C.O. Impact of urbanization and socio-economic status on infant feeding practices in Lagos, Nigeria. *Food and Nutr. Bull.* 1985; 7:33–37.

44. Ziegler, J.B., Cooper, D.A., Johnson, R.O., and Gold, J. Postnatal transmission of AIDS-associated retrovirus from mother to infant. *Lancet* 1985; 1:896–98.

45. Newburg, D.S., Viscidi, R.P., Ruff, A., and Yolken, R.H. A milk factor inhibits binding of human immunodeficiency virus to the CD_4 receptor. *Pediat. Res.* 1992; 31:22-28.

46. Dunn, D.T., Newell, M.L., Ades, A.E., and Peckham, C.S. Risk of human immunodeficiency virus type 1 transmission through breastfeeding. *Lancet* 1992; 340:585–588.

47. WHO/UNICEF. Breast-feeding and HIV. WHO Press Release, No. 30, 4 May 1992. In: *Progress in Human Reproduction Research* No. 23, pp. 6–7, 1992; World Health Organization, Geneva.

48. Nicoll, A., Killewo, J.Z.J., and Mgone, C. HIV and infant feeding practices: epidemiological implications for sub-Saharan African countries. *AIDS* 1990; 4:661–665.

49. Fox, P.C. Saliva and salivary gland alterations in HIV infection. *JADA* 1991; 122:46–48.

50. Fox, P.C., Wolff, A., Yeh, C.K., Atkinson, J.C., and Baum, B.J. Saliva inhibits HIV-1 infectivity. *JADA* 1988; 116:635–637.

51. Editorial. Structural adjustment and health in Africa. *Lancet* 1990; 335:885–86.
52. World Bank. Sub-Saharan Africa: from crisis to sustainable growth. A long term perspective study. Washington, D.C.: World Bank, 1989.
53. Bell, D. and Michael R. Health, *Nutrition and Economic Crises: Approach to Policy in the Third World*. Dover, MA: Auburn House, 1988.
54. Editorial, Poverty, malnutrition and world food supplies. *Lancet* 1987; I:487–488.
55. Maletnlema, T.N. The problem of food and nutrition in Africa. *World Rev. Nutr. Diet* 1986; 47:30–79.
56. Brown, K.H. and Solomons, N.W. Nutritional problems of developing countries. *Infect. Dis. Clin. N. Am.* 1992; 5:297-317.
57. Keusch, G.T. and Farthing, M.J.G. Nutrition and infection. *Annu. Rev. Nutr.* 1986; 6:131-54.
58. Schuerman, L., Seynhaeve, V., Bachschmidt, I., Tchotch, V., Quattara, S.A., and de The, G. Severe malnutrition and pediatric AIDS: a diagnostic problem in rural Africa. *AIDS* 1988; 2:232–233.
59. Moseson, M., Zeleniuch-Jacquotte, A., Belsito, D.V., Shore, R.E., Marmon, M., and Pasternack, B. The potential role of nutritional factors in the induction of immunologic abnormalities in HIV-positive homosexual men. *J. Acquir. Immun. Def. Synd.* 1989; 2:235–247.
60. Chlebowski, R.T. Significance of altered nutritional status in acquired immune deficiency syndrome (AIDS). *Nutr. Can.* 1985; 1-2:85-91.
61. Stein, T.P., Nutinsky, C., Condoluci, D., Schluter, M.D., and Leskiw, M.J. Protein and energy substrate metabolism in AIDS patients. *Metabolism* 1990; 39:876–881.
62. Kottler, D.P. Nutritional support in AIDS. *Am. J. Gastroenterol.* 1990; 86:539–541.
63. Grunfeld, C., Pang, M., Shimizu, L., Shigenaga, J.K., Jensen, P., and Feingold, K.R. Resting energy expenditure, caloric intake, and short-term weight change in human immunodeficiency virus infection and the acquired immunodeficiency syndrome. *Am. J. Clin. Nutr.* 1992; 55:455–460.
64. Ott, M., Lembeke, B., Fischer, H., Jager, R., Polat, H., Geier, H., Rech, M., Staszewski, S., Helm, E.B., and Caspary, W.F. Early changes of body composition in human immunodeficiency virus-infected patients:tetrapolar body impedance analysis indicates significant malnutrition. *Am. J. Clin. Nutr.* 1993; 57:15–19.
65. Baker, D.H. and Wood, R.J. Cellular antioxidant status and human immunodeficiency virus replication. *Nutr. Rev.* 1992; 50:15–18.
66. Beach, R.S., Mantero-Atienza, E., Shor-Posner, G., Javier, J.J., Szapocznik, J., Morgan, R., Sauberlich, H.E., Cornwell, P.E., Eisdorfer, C., and Baum, M.K. Specific nutrient abnormalities in asymptomatic HIV-1 infection. *AIDS* 1992; 6:701–708.
67. Boudes, P., Zittoun, J., and Sobel, A. Folate, vitamin B_{12}, and HIV infection. *Lancet* 1990; 355:1401–1402.
68. Ward, B.J., Humphrey, J.H., Clement, L., and Chaisson, R.E. Vitamin A status in HIV infection. *Nutr. Res.* 1993; 13:157–166.
69. Buhl, R., Holroyd, K.J., Mastrangeli, A., Cantin, A.M., Jaffe, H.A., Wells, F.B., Saltini, C., and Crystal, R.G. Systemic glutathione deficiency in symptom-free HIV-seropositive individuals. *Lancet* 1989; 2:1294–98.
70. Meister, A. *Glutathione centennial: molecular properties and clinical implications*. New York: Academic Press, 1989; 3–21.

71. de Quay, B., Malinverni, R., and Lauterburg, B.H. Glutathione depletion in HIV-infected patients: role of cysteine deficiency and effect of oral N-acetylcysteine. *AIDS* 1992; 6:815–819.

72. Gmunder, H., Eck, H.P., Benninghoff, B., Roth, S. and Droge, W. Macrophages regulate intracellular glutathione levels in lymphocytes. *Cell Immunol.* 1990; 129:32–46.

73. Jackson, A.A. Blood glutathione in severe malnutrition in childhood. *Trans. R. Soc. Trop. Med. Hyg.* 1986; 80:911–913.

74. Wiezbicker, G.T., Hegen, T.M., and Jones, D.P. Glutathione in food. *J. Food Composit.* 1989; 2:327–337.

75. Keusch, G.T. and Farthing, M.J.F. Nutritional aspects of AIDS. *Annu. Rev. Nutr.* 1990; 10:475–501.

76. Kottler, D.P. Nutritional effects and support in the patient with acquired immunodeficiency syndrome. *J. Nutr.* 1992; 122:723–727.

77. Grimble, R.F. Nutrition and cytokine action. *Nutr. Res. Rev.* 1990; 3:193–210.

78. Grimble, R.F. Dietary manipulation of the inflammatory response. *Proc. Nutr. Soc.* 1992; 51:285–294.

79. Murray, H.W., Rubin, B.Y., Masur, H., and Roberts, R.B. Impaired production of lymphokines and immune (gamma) interferon in the acquired immunodeficiency syndrome. *New Engl. J. Med.* 1984; 310:883–889.

80. Lahdevirta, J., Maury, C.P., Teppo, A.M., and Repo, H. Elevated levels of circulating cachectin/tumor necrosis factor in patients with acquired immunodeficiency syndrome. *Am. J. Med.* 1988; 85:289–91.

81. Matsuyama, T., Kobayashi, N., and Yamamoto, N. Cytokines and HIV infection: Is AIDS a tumor necrosis factor disease? *AIDS* 1991; 5:1405–1407.

82. Breen, E.C., Rezai, A.R., Nakajirma, K., et al. Infection with HIV is associated with elevated IL-6 levels and production. *J. Immunol.* 1990; 144:480–484.

83. Chaisson, R.E. and Volberding, P.A. Clinical manifestations of HIV infection. In: Mandell, G.L., Douglas, R.G., and Bennett, J.E. (eds). *Principles and Practice of Infectious Diseases.* New York: Churchill Livingstone, 1990; 1059–1091.

84. Carman, J.A and Hayes, C.E. Abnormal regulation of IFN-γ secretion in vitamin A deficiency. *J. Immunol.* 1991; 147:1247–52.

85. Abril, E.R., Rybski, J.A., Scuderi, P., and Watson, R.R. Beta-carotene stimulates human leukocytes to secret a novel cytokine. *J. Leukocyte Biol.* 1989; 45:255–261.

86. Edelman, R. Cell-mediated immune response in protein-calorie malnutrition. A review. In: Suskind, R.M. (ed.) *Malnutrition and the Immune Response.* Raven Press, New York, 1977; 47–67.

87. Griffin, G.E., Leung, K., Folks, T.M., Kunkel, S., and Nabel, G.J. Activation of HIV gene expression during monocyte differentiation by induction of NF-Kappa B. *Nature* 1989; 339:70–73.

88. Poli, G., Bressler, P., Kinter, A., et al. Interleukin 6 induces human immunodeficiency virus expression in infected monocytic cells alone and in synergy with tumor necrosis factor alpha by transcriptional and post-transcriptional mechanisms. *J. Exp. Med.* 1990; 172:151–158.

89. Karin, M. Signal transduction from cell surface to nucleus in development and disease. *FASEB J.* 1992; 6:2581–90.

90. Riviere, Y., Blank, V., Kourilsky, P., and Israel, A. Processing of the precursor of NF-kB by the HIV-1 protease during acute infection. *Nature* 1991; 350:625–626.

91. Kolberg, B. Basic and clinical HIV-1 researchers target NF-kB protein. *J. NIH Research* 1991; 3:28–29.
92. Kalebic, T., Kinter, A., Poli, G., Anderson, M.E., Meister, A. and Fauci, A.S. Suppression of human immunodeficiency virus expression in chronically infected monocytic cells by glutathione, glutathione ester, and N-acetylcysteine. *Proc. Natl. Acad. Sci. (USA)* 1991; 88:986–990.
93. Enwonwu, C.O. Experimental protein-calorie malnutrition in the guinea pig and evaluation of the role of ascorbic acid status. *Lab. Invest.* 1973; 29:17–26.
94. Bowman, T.A., Goonewardene, I.M., Pasatiempo, A.M., Ross, A.C., and Taylor, C.E. Vitamin A deficiency decreases natural killer cell activity and interferon production in rats. *J. Nutr.* 1990; 120:1264–1273.
95. Bogden, J.D., Baker, H., Frank, O., Perez, G., Kemp, F., Bruening, K., and Louria, D. Micronutrient status and human immunodeficiency virus (HIV) infection. *Ann. NY Acad. Sci.* 1990; 587:189–195.
96. Harakeh, S., Jariwalla, R.J., and Pauling, L. Suppression of human immunodeficiency virus replication by ascorbate in chronically and acutely infected cells. *Proc. Natl. Acad Sci. (USA)* 1990; 87:1745–49.
97. Broquist, H.P. Buthionine Sulfoximine, an experimental tool to induce glutathione deficiency: elucidation of glutathione and ascorbate in their role as antioxidants. *Nutr. Rev.* 1991; 50:110–111.
98. Deo, G.M. Cell biology of protein-calorie malnutrition. *World Rev. Nutr. Diet* 1978; 32:49–95.
99. Samaranayake, L.P. Nutritional factors and oral candidosis. *J. Oral Pathol.* 1986; 15:61–65.
100. Federation Dentaire Internationale. Saliva: its role in health and disease. *Int. Dent. J.* No. 4, Suppl. 2, 1992; 42:291–304.
101. Mandel, I.D. The function of saliva. *J. Dent. Res.* 1987; 66:623–627.
102. Hickey. M.S. Nutritional support of patients with AIDS. *Surg. Clin. N. Am.* 1991; 71:645–664.
103. Winick, M. The National Task Force on Nutrition in AIDS. Guidelines on nutritional support in AIDS. *Nutrition* 1989; 5:390–394.

Chapter

3

AIDS in Europe

Diego Serraino,

Silvia Franceschi,

and

Carlo La Vecchia

SUMMARY

By the end of March 1992, nearly 70,000 cases of acquired immunodeficiency syndrome (AIDS) have been reported to the World Health Organization (WHO) by 19 countries of the WHO European region. Most of the cases (67,322/69,682, 97%) were recorded in Western Europe, chiefly in five countries: France (18,926 cases), Spain (13,261), Italy (12,754), Germany (7,957) and the United Kingdom (5,782). Since the report of the first European cases of AIDS in 1981 until 1987, AIDS spread faster in the Northern and in the Central parts than elsewhere in Europe. Thereafter, the growth of the epidemic has been markedly more rapid in Southern Europe, while in Eastern Europe AIDS is still in an early phase. Nearly 5% of the total number of European AIDS cases occurred in children (less than 13 years of age) and most of them (75%) were due to iatrogenically acquired human immunodeficiency virus (HIV) infection in Romania. Children with AIDS in Western European countries, on the contrary, acquired HIV infection from their mother in approximately 80% of the cases. Of all adults with AIDS, 9,388 (14%) were women. The majority of them (56%) were intravenous drug users (IVDU), in particular in Southern Europe, while the second largest group (29% of cases) was represented by heterosexual women. In the last years of the epidemic, the percentage of IVDU women with AIDS has declined, whereas the proportion

0-8493-7842-7/94/$0.00+$.50
© 1994 by CRC Press, Inc.

of heterosexual women increased (from about 20% before 1987 to 30% in 1991). Eighty-five percent of adult cases in Europe were among men. Nearly 50% of such cases occurred in homosexual or bisexual men (28,382 out of 57,134) and 32% in IVDUs. Whereas AIDS cases in homosexuals or bisexuals were concentrated in Northern (79%) or in Central Europe (59%), the largest proportion of IVDUs with AIDS was registered in Southern Europe (65%). Over time, the yearly number of new cases in IVDU men nearly equalized that registered in homosexual or bisexual men (4,594 and 5,064 in 1991, respectively). The percentage of AIDS cases registered among IVDU men sharply increased from 13% before 1986 to 40% in 1991, whereas among homosexual or bisexual men it decreased from 69% to 44%.

INTRODUCTION

Worldwide, three major epidemiologic patterns of AIDS have been described.[1] In Pattern I, the vast majority of AIDS cases are recorded in homosexual or bisexual men and IVDUs, their sexual partners and offspring. Such a pattern occurs in North America, Western Europe and Australia. Pattern II includes sub-Sahara African countries and other parts of the world, like South America, where infection with human immunodeficiency virus (HIV or HIV-1) is mainly transmitted heterosexually, perinatally or by blood transfusion. Pattern III occurs in those countries where the AIDS epidemic is still in a very early phase and where a small number of cases infected in various ways has been reported, such as Eastern European countries.[1]

This contribution outlines the main features of the epidemiology of AIDS in Europe, since the beginning of the epidemic through March 1992, with particular regard to geographic and temporal patterns in children and adults.

The 31st March 1992 update of the European Non Aggregate AIDS Data Set was analyzed.[2] Such data set contains information for each individual AIDS case recorded by the national surveillance system of 19 countries of the WHO European Region (Table 1). The data are collected by the WHO Collaborating Centre on AIDS in Paris, France, according to a standard core of epidemiologic information which include, for each case, country of report, sex, age, year of AIDS diagnosis, HIV transmission category (according to the Centers for Disease Control [CDC] criteria)[3,4] and reported indicator disease at the time of AIDS diagnosis.

GEOGRAPHIC PATTERNS

There is consistent evidence that HIV infection was first introduced in Europe by subjects who had heterosexual contacts in Africa,[5] and, mostly, by homosexual men who were sexual partners of infected men from the United

States.[6,7] From the report of the first cases through 1987, AIDS spread faster in Northern and Central Europe than elsewhere (Figure 1). Thereafter, the growth of the epidemic has been markedly more rapid in the South than in the rest of Western Europe, whereas in Eastern European countries the AIDS epidemic is still in a very early phase (Figure 1).

By the 31st of March 1992, a cumulative number of 69,682 AIDS cases has been reported to the WHO by 19 countries of the European region (Table 1). Switzerland, Spain and France have shown the highest incidence rates in Western Europe (from 70 to 80 cases per million population), while the lowest rates were registered in Greece and Israel (with less than ten cases per million population).[2] Countries in the eastern part of the region reported, generally, very low numbers of cases, with rates below 3 cases per million population. An exception is constituted by Romania, where nearly 1,500 pediatric cases have been reported in the period 1990–91.[2]

A considerable between-country variation exists in the proportion of cases recorded across HIV transmission categories. Most of the 28,382 cases recorded in homosexual or bisexual men were from Northern (12,232 cases, 43%) and Central Europe (11,732 cases, 41%), in particular from Germany (5,563), United Kingdom (4,496), the Netherlands (1,640) and France (9,940) (Table 1). On the other hand, 13,743 out of 23,940 cases (71%) that occurred in IVDUs were registered in Southern Europe, while heterosexual cases were concentrated in Central Europe (50%) (Table 1).

AIDS IN CHILDREN

Approximately 5% (3,155 out of 69,682 cases) of AIDS cases reported in the WHO European Region were registered in children (12 years of age or younger).[2] The vast majority of pediatric cases (75%) were concentrated in Eastern Europe, while in Western Europe children with AIDS accounted for 1% of total cases in the Northern and for 3% of cases in the Southern parts (Table 2). Such a difference was largely explained by an outbreak of AIDS in Romania that started in 1989 (Figure 2) among abandoned children living in public institutions.[8] Only 8% of the mothers of Romanian children with AIDS were known to be seropositive for HIV antibodies, and it was ascertained that at least 40% of cases were in children who had received transfusions of unscreened blood. Moreover, the acquisition of HIV infection through the improper use of needles and syringes was strongly suspected in the majority of the children in whom the mode of HIV infection was not known.[8]

Most of the pediatric cases in Western Europe, on the contrary, acquired HIV infection from an infected mother, with percentages ranging from 71% in Northern Europe to 85% in Southern Europe (Table 2). Seventy-five percent of pediatric cases in Western Europe were registered in three countries, Spain (415 cases), France (401) and Italy (283),[2] thus reflecting the

TABLE 1.
Distribution of AIDS Cases in Adults by HIV Transmission Category. Europe, 1981–1992

| | AIDS Cases | | HIV Transmission Category | | | | | | |
	Men (57,134) No.	Women (9,388) No.	Homosexual or Bisexual Men %	Intravenous Drug Users Men %	Women %	Heterosexuals Men %	Women %	Recipients of Blood or Blood Products Men %	Women %
Northern Europe									
Germany	7,281	614	76	10	53	3	28	5	11
Iceland	19	3	95	0	33	0	0	0	67
Netherlands	1,920	137	85	7	39	5	41	2	10
Sweden	614	54	76	6	18	6	43	8	22
United Kingdom	5,377	327	82	5	23	6	59	6	14
Total	15,211	1,135	79	8	41	4	39	5	13
Central Europe									
Austria	618	113	50	26	49	6	34	7	11
Belgium	809	228	52	6	7	37	74	4	17
France	15,807	2,731	61	21	39	8	36	5	17
Switzerland	1,894	470	53	33	63	10	31	2	4
Total	19,128	3,542	59	22	41	9	37	5	15

Southern Europe									
Greece	488	58	59	4	12	17	41	12	28
Israel	164	9	52	18	44	7	44	17	11
Italy	10,146	2,335	19	69	71	4	21	3	3
Portugal	755	104	51	14	24	22	56	7	14
Spain	10,760	2,102	19	68	76	4	16	4	3
Total	22,313	4,608	21	65	71	5	20	4	4
Eastern Europe									
Czechoslovakia	27	1	85	0	0	0	0	7	0
Hungary	77	4	75	1	0	4	50	17	25
Poland	92	11	58	33	73	9	27	0	0
Romania	63	47	18	0	0	24	25	21	19
Yugoslavia	223	40	23	43	55	11	30	17	10
Total	482	103	41	26	30	11	28	14	14

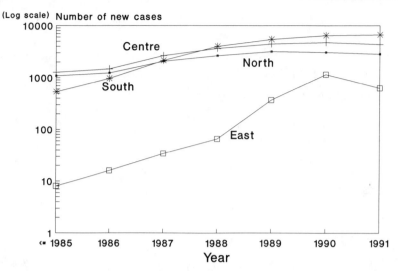

Figure 1.   The epidemic curves of AIDS cases in various areas. Europe, 1981–1992.

greater involvement of women in the AIDS epidemic in Southern Europe. Whereas in France 54% of the infected mothers acquired HIV infection by heterosexual contacts, in Spain and Italy the infected mothers were IVDUs in 60% of the cases.[2]

The assessment of the risks of HIV transmission from mother to newborn is difficult, partly because of the persistence for up to one year of maternal antibodies passively transmitted to the infant. Approximately 1 out of 3 babies born to HIV-infected mothers will have evidence of infection or of AIDS by the age of 18 months, and about one fifth of them will have died.[9,10] The precise mechanism of mother-to-newborn transmission of HIV infection is not clear yet. HIV transmission from mother to child may be transplacental, or attributable to the admixture of maternal and fetal blood during placental separation or consequent to the passage through the birth canal.[9]

Prevention of HIV transmission from mother to child in Europe, particularly in Western countries, must be focused on encouraging women who are IVDUs, or partners of IVDUs or of high-risk subjects, to undergo HIV testing and counseling. Infected women should be convinced to use effective contraception. Furthermore, on account of the possibility, albeit not well quantified, of HIV transmission through lactation, HIV-positive women should be advised to abstain from breast-feeding their babies. This advice is at variance with the one given in the poor areas of the Third World, where mother-to-child transmission of HIV infection is common, but the benefits of breast-feeding are thought to outweigh the risks of HIV transmission.

On the other hand, the outbreak of the AIDS epidemic in children in Romania demonstrates the serious potential for HIV transmission in medical facilities in countries with poorly organized and/or underfunded health systems.

TABLE 2.
Distribution of Pediatric AIDS Cases, by Transmission Category and
Geographic Area. Europe, 1981–1992

	Pediatric Cases (3,155) No.	Percent of All Cases %	HIV Transmission Category	
			Mother to Newborn %	Recipient of Blood or Blood Products %
Geographic area:				
Northern Europe	166	1	71	29
Central Europe	509	2	81	18
Southern Europe	705	3	85	14
Eastern Europe*	1,775	75	6	32

* Information on HIV transmission category was not available for 62% of cases.

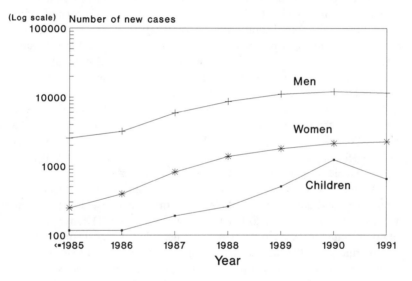

Figure 2.   The epidemic curves of AIDS cases in children and adults. Europe, 1981–1992.

AIDS IN WOMEN

In Western countries women account for an increasing number and percentage of adults with AIDS.[11] Since the report of the first cases of AIDS among European women, registered in Northern Europe in the early 1980's, more than 9,000 cases occurred as of March 1992 (Figure 2), representing 14% of the cumulative number of AIDS cases in European adults (Table 1). The largest proportion of women with AIDS was among IVDUs (5,220 out of 9,388 cases, 56%), in particular in Southern Europe where 71% of cases were attributable to intravenous use of drugs. The second largest group (29% of total cases) was represented by women who acquired HIV infection through sexual intercourse with an infected male partner (Table 1). At the time of AIDS diagnosis,[2] 53% of women with AIDS were in their 20s and 90% were of reproductive age (15 to 44 years).

Although intravenous use of drugs is by far the most important risk factor for HIV infection among European women, it is interesting to note that the number of AIDS cases reported each year among heterosexual women steadily rose from 1989 onward, whereas it appeared to level off among IVDUs (Figure 3). As a proportion of AIDS cases, the percentage of IVDU women started to decline after 1988 (when 60% of European AIDS cases were IVDUs), whereas the proportion of heterosexual women sharply increased up to 31% in 1991 (Figure 4).

Overall, 9% of adult women with AIDS acquired HIV infection through blood transfusion. More than 60% of European cases in such at-risk groups were reported in Central Europe, in particular in France and in Belgium where 17% of all women with AIDS were infected through transfusions of infected blood (Table 1). The absolute number of yearly new cases attributable to transfusion of blood, or of blood derivatives, increased up to 1989 (Figure 3). Since the mean interval between contamination and AIDS is about 8 years,[12] the effects of measures taken in 1985–1986 to prevent HIV transmission through blood transfusions could not yet be observed.

As a consequence of the increasing involvement of women in the epidemic, the heterosexual transmission of HIV has generated much interest and debate. Studies based on heterosexual partners of persons infected with HIV have consistently documented the occurrence of male-to-female and, to a lesser extent, female-to-male sexual transmission of HIV infection.[13-16]

A cross-sectional study on 153 couples from Greece, France, Spain, Netherlands and Italy, in which the male partner was infected identified three major risk factors for male-to-female HIV transmission: (1) a history of sexually transmitted diseases in the female partner; (2) a diagnosis of AIDS in the male partner; and (3) the practice of anal intercourse.[13] Among 368 Italian women who were steady partners of HIV-infected men, an increased risk of acquiring HIV infection was found among women who reported the practice of anal intercourse and in those who had a history of sexually transmitted diseases.[17]

The importance of preventing the heterosexual transmission of HIV infection for limiting the spread of the epidemic in the general population could not be overemphasized. Since several behavioral and biologic factors that increase the risk of HIV transmission are already well known, all sexually active women should be made aware of the potential of acquiring HIV infection and AIDS from sexual activity. Beside a stable, mutually monogamous relationship with an uninfected partner, which eliminates any new risk of sexually transmitted HIV infection, it should be stressed that the reduction of the number of sexual partners, the avoidance of anal intercourse and, most of all, the consistent use of condoms are the most important means to prevent HIV infection.[18]

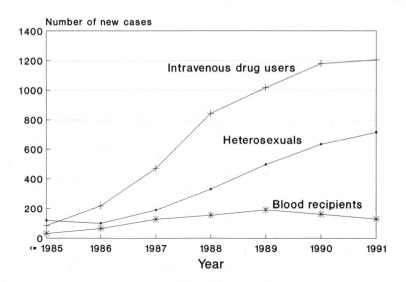

Figure 3.   Yearly number of new cases in women, by HIV transmission category. Europe, 1981–1992.

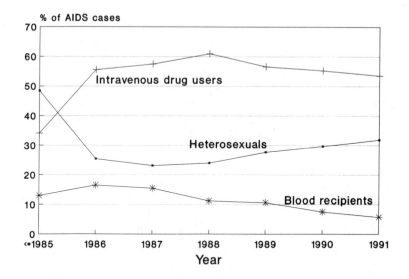

Figure 4.   Percentage of women with AIDS by year of diagnosis and HIV transmission category. Europe, 1981–1992.

AIDS IN MEN

Approximately 85% of all cases of AIDS reported in adults in Europe occurred among men (Table 1). The number of new cases registered each year among men has been steadily increasing since the beginning of the epidemic (from 2,540 before 1986 to 11,558 in 1991) (Figure 2).

Of the 57,134 cases registered among adult men, nearly half (28,382) occurred in homosexuals or bisexuals. The risk of HIV infection among these men substantially increases with the number of sexual partners, the frequency of receptive anal intercourse and of other practices that can result in rectal trauma.[6,19-22] The number of new cases documented each year among homosexual or bisexual men has greatly exceeded the number recorded in any other at-risk group up to 1989. Thereafter, the number of AIDS cases among IVDU men nearly equalized that registered among homosexual or bisexual men (4,594 and 5,064, respectively, in 1991) (Figure 5).

The change in the contribution of homosexual or bisexual men to the growth of the AIDS epidemic in Europe is also well documented by the sharp decline in the proportion of total AIDS cases diagnosed in such groups, from 69% before 1985 to 44% in 1991 (Figure 6). In the United States, the proportion of cases recorded among homosexual or bisexual men has also decreased over time, but less markedly than in Europe, from 63% before 1985 to 59% in 1991.[23,24]

Within 2 to 3 years since the report of the first AIDS cases in homosexual or bisexual men, HIV infection rapidly involved IVDUs who lived in some metropolitan areas of Europe, particularly in the Northern part of Italy,[25-28] in Spain[29] and in Scotland.[30] The sharing of injection equipment among IVDUs was quickly identified as the most important mode of HIV transmission.[31] Interestingly, needle sharing is very common among IVDUs also in those countries, like Italy, where sterile needles and syringes are inexpensive and legally available without medical prescription.[25]

In Europe, 32% of all AIDS cases reported as of March 1992 in adult men were in heterosexual IVDUs. In such groups, AIDS grew faster than in any other transmission category, both in absolute terms (Figure 5) and as a percentage of total cases (Figure 6). The overall proportion of cases diagnosed among IVDUs has increased sharply from 13% before 1986 to 40% in 1991.

A well-defined north-south geographic gradient in the distribution of AIDS cases among IVDUs in Europe has been identified since the mid 1980's. Of the 19,938 cases diagnosed in IVDUs as of March 1992, 14,465 (73%) were concentrated in Southern Europe, particularly in Italy (10,146 cases) and Spain (7,273 cases). In these two countries, nearly 70% of the national cases were reported in IVDUs (Table 1). In the other areas, IVDUs accounted for 8% of AIDS cases in Northern Europe and 22% in Central Europe. In the Eastern part of Europe, Poland (30 cases) and the former Yugoslavia (96 cases) were the only countries to report AIDS cases in IVDU men (Table 1).

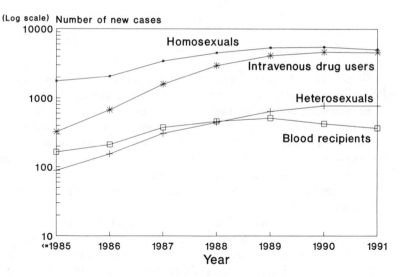

Figure 5. Yearly number of new cases in men, by HIV transmission category. Europe, 1981–1992.

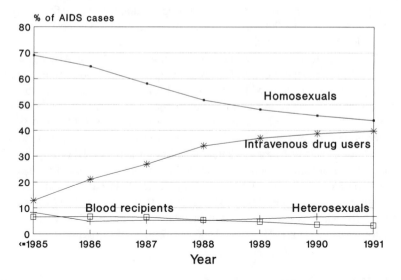

Figure 6. Percentage of men with AIDS by year of diagnosis and HIV transmission category. Europe, 1981–1992.

Since many more men than women are infected with HIV, sexual transmission of HIV from infected women to their male partners has been poorly studied. An European multicentre study based on 156 female index patients and their 159 male partners found that female-to-male transmission of HIV is about 50% less effective than the male-to-female transmission.[14] Age and stage of infection of the female partner and the practice of anal intercourse were found to increase the risk of male-to-female transmission of HIV infection.[14]

Six percent of European men with AIDS (3,458 out of 57,134) acquired HIV infection through sexual contacts with infected women (Table 1). The highest proportion of heterosexually acquired cases was registered in Central Europe (9% of cases), in particular in Belgian men (37% of AIDS cases) (Table 1) who have chiefly acquired HIV infection through sexual contacts with infected prostitutes in Africa.[5] As previously described in women, also the contribution of heterosexual men to the growth of the epidemic substantially increased over time, both in absolute terms (Figure 5) and as a proportion of total cases (Figure 6).

Although in most Western countries the risk of transmission of HIV infection by transfusion of blood or blood products has been considerably reduced,[12] HIV antibody screening of blood donations for the period 1988–1990 still shows relatively high rates of infection (>1 per 10,000 donations) in some European countries like France, Portugal, Spain and Romania. In all these countries, rates of HIV-positive donations were found to be much higher in first-time donors (3.0/10,000) than in regular donors (0.2/10,000).[32] It is estimated that at least 90% of recipients of an HIV-antibody positive blood transfusion become infected.[12]

Approximately 5% of the cumulative number of cases of AIDS in adult men in Europe have occurred among recipients of blood transfusions, haemophiliacs or persons with coagulation disorders (2,592 cases) (Table 1). In Western European countries, the proportion of cases diagnosed in recipients of blood transfusions or blood products was less than 5% (Table 1). A much higher proportion of iatrogenically acquired cases of AIDS has been reported in the Eastern part of Europe, where 14% of cases acquired HIV infection by transfusion of infected blood or blood products (Table 1). Most of such cases were recorded in Romania and Yugoslavia (Table 1). The absolute number of new cases recorded each year in blood recipients steadily rose up to 1988–89 (Figure 5), whereas, as a proportion of total AIDS cases, it showed a clear downward trend (from 7% before 1986 to 3% in 1991) (Figure 6).

CONCLUSIVE REMARKS

Few points can be addressed to summarize the major epidemiologic aspects of AIDS in Europe. First, it is becoming increasingly feared that heterosexual IVDUs will be, in the short term, the population group most heavily affected by the AIDS epidemic. This fact has important implications for the prevention and control of the epidemic in Western Europe, in the light of the well-documented heterosexual transmission of HIV infection from IVDUs to the general population. Therefore, programs aimed at reducing the spread of HIV infection should be chiefly targeted to IVDUs in order to be most effective, in particular in those countries, like Italy and Spain, where the vast majority of cases are attributable to drug use. Preventive strategies should include detoxification programs as well as the availability of sterile needles, syringes and condoms.

The second important topic concerns women. The increasing number of European women with AIDS mirrors — with a delay of approximately four years — the picture from the United States, where the proportion of women with heterosexually acquired AIDS tends to increase more rapidly than that of IVDU women. Several factors that potentiate the risk of male-to-female sexual transmission of HIV infection have been identified and such knowledge should help addressing effective preventive measures, first of all condom use.

A last comment pertains to the AIDS epidemic in Eastern European countries. So far, the limited interaction between these countries and other European areas may have delayed or slowed the spread of HIV into these nations. Such delay can theoretically offer the opportunity to take appropriate preventive steps when they are most effective (i.e., in the early phase of the epidemic). Furthermore, the potential for HIV transmission in medical facilities seems to be of an unprecedented importance in Eastern Europe and requires immediate action.

ACKNOWLEDGMENTS

Data from the European Non-Aggregate AIDS Data Set (ENAADS), version AIDS9203.DAT, prepared by the European Centre for the Epidemiological Monitoring of AIDS in Paris, France were used in this study. Compilation of this data file was made possible by the continuing participation of clinical doctors in mandatory and voluntary national schemes for AIDS cases.

This work was supported by two grants from "Ministero della Sanità-Instituto Superiore di Sanità, V Progetto AIDS, 1992, contracts No. 7201-29/7203-03", Rome, Italy.

REFERENCES

1. Piot P, Plummer FA, Mhalu FS, et al. AIDS: an international perspective. *Science* 1988; 239:573–579.
2. European Centre for the Epidemiological Monitoring of AIDS: AIDS Surveillance in Europe. Quarterly Report N. 33, 31st March 1992.
3. Centers for Disease Control. Revision of the CDC surveillance case definition for acquired immunodeficiency syndrome. *MMWR* 1985; 34:374–375.

4. Centers for Disease Control. Revision of the CDC surveillance case definition for the acquired immunodeficiency syndrome. *MMWR* 1987; 36:3s-15s.
5. Bonneux L, Van der Stuyft P, Taelman H, et al. Risk factors for infection with human immunodeficiency virus among European expatriates in Africa. *Br Med J* 1988; 297:581–584.
6. Melbye M, Biggar RJ, Ebbesen P, et al. Seroepidemiology of HTLV-III antibody in Danish homosexual men: prevalence, transmission and disease outcome. *Br Med J* 1984, 289:573–575.
7. Serraino D, Franceschi S, Vaccher E, et al. Geographic factors and human immunodeficiency virus infection among homosexual men and intravenous drug abusers from the northeastern part of Italy. In: Fleming A, Carballo M, Fitz-Simons DW, et al (eds.) *The Global Impact of AIDS*. New York: Alan R. Liss Inc., 1988. pp. 43–52.
8. Hersh BS, Popovici F, Apetrei RC, et al. Acquired immunodeficiency syndrome in Romania. *Lancet* 1991; 338:645–649.
9. Boylan L and Stein ZA. The epidemiology of HIV infection in children and their mothers — vertical transmission. *Epidemiol Rev* 1991; 13:143–177.
10. European Collaborative Study. Children born to women with HIV-1 infection: natural history and risk of transmission. *Lancet* 1991; 337: 253–260.
11. Ellerbrock TV, Bush TJ, Chamberland ME, and Oxtoby MJ. Epidemiology of women with AIDS in the United States, 1981 through 1990. *JAMA* 1991; 265:2971–2975.
12. Goedert JJ and Blattner WA. The epidemiology and natural history of human immunodeficiency virus and AIDS, in De Vita Jr VT, Hellman S, Rosenberg SA (eds.): *AIDS, Etiology, Diagnosis, Treatment and Prevention*. Philadelphia, PA, Lippincot, 1985, pp 3–60.
13. European Study Group. Risk factors for male to female transmission of HIV. *Br Med J* 1989; 298:411–415.
14. European Study Group. Comparison of female to male and male to female transmission of HIV in 563 stable couples. *Br Med J* 1992; 304:809–813.
15. Padian N, Marquis L, Francis DP, et al. Male-to-female transmission of human immunodeficiency virus. *JAMA* 1987; 258:788–790.
16. Padian NS, Shiboski SC, and Jewell NP. Female-to-male transmission of human immunodeficiency virus. *JAMA* 1991; 266:1664–1667.
17. Lazzarin A, Saracco A, Musicco M, et al. Man-to-woman sexual transmission of the human immunodeficiency virus. *Arch Intern Med* 1991; 151:2411–2416.
18. Haverkos HW and Edelman R. The epidemiology of acquired immunodeficiency syndrome among heterosexuals. *JAMA* 1988; 260: 1922–1929.
19. Curran JW, Jaffe HW, Hardy AM, et al. Epidemiology of HIV infection and AIDS in the United States. *Science* 1988; 239:610–616.
20. Sirianni MC, Rossi P, Moroni M, et al. Demonstration of antibodies to human T-lymphotropic retrovirus type III in lymphoadenopathy syndrome patients and in individuals at risk for acquired immunodeficiency syndrome (AIDS) in Italy. *Am J Epidemiol* 1986; 123:308–315.
21. Franceschi S, Serraino D, Saracchini S, et al. Risk factors for human immuno-deficiency virus (HIV) infection among homosexual and bisexual men in a region at low risk for AIDS: the Northeastern part of Italy. *Rev Epidem et de Sante Publ* 1989; 37:103–108.

22. Griensven Van GJP, Tielman RAP, Goudsmit J, et al. Risk factors and prevalence of HIV antibodies in homosexual men in the Netherlands. *Am J Epidemiol* 1987; 125:1048–57.
23. Center for Disease Control Statistics. *AIDS* 1991; 5:1401–1403.
24. Berkelman RL and Curran JW. Updates: Epidemiology of HIV infection and AIDS. *Epidemiol Rev* 1989; 11:222–228.
25. Serraino D, Franceschi S, Vaccher E, et al. Risk factors for human immunodeficiency virus infection in 581 intravenous drug users, Northeast Italy, 1984–1988. *Int J Epidemiol* 1991; 20:264–270.
26. Franceschi S, Tirelli U, Vaccher E, et al. Risk factors for HIV infection in drug addicts from the Northeast Italy. *Int J Epidemiol* 1988; 17:162–167.
27. Aiuti F, Rossi P, Sirianni MC, et al. IgM and IgG antibodies to human T cell lymphotropic retrovirus (HIV) in lymphadenopathy syndrome and subjects at risk for AIDS in Italy. *Br Med J* 1985; 291: 165–166.
28. Ferroni P, Geroldi D, Galli C, et al. HIV antibody among Italian drug addicts (Letter). *Lancet* 1985; 2:52–53.
29. Rodrigo JM, Serra MA, Aguilar E, et al. HTLV–III antibodies in drug addicts in Spain (Letter). *Lancet* 1985; 2:156–157.
30. Follett EAC, McIntyre A, O'Donnel B, et al. HTLV-III antibody in drug abusers in the east of Scotland: The Edinburgh connection (Letter). *Lancet* 1986; 1:446–447.
31. Des Jarlais DC and Friedman SR. Editorial review: HIV infection among intravenous drug users: epidemiology and risk reduction. *AIDS* 1987; 1:67–76.
32. European Centre for the Epidemiological Monitoring of AIDS: AIDS Surveillance in Europe, Quarterly Report N. 30, 30th June 1991.

Nutrition In HIV-Positive Drug Addicts

Pilar Varela

and

Ascension Marcos

As of July 1, 1992, a cumulative total of 501,272 cases of acquired immunodeficiency syndrome (AIDS) has been reported to WHO from 191 countries/areas. However, WHO estimates that about 2 million people, including more than a half million children, have developed AIDS since the beginning of the pandemic. In order to have a realistic picture of the pandemic today, one must look not at AIDS cases but at the number of people infected with HIV. WHO estimates that at least 10-12 million adults and children have been infected with HIV to date.[1] Within this context, the second largest group of cases with AIDS in USA[2] and Europe[1] has occurred in persons who use illegal drugs intravenously (i.v.). HIV parenteral transmission also is the second major mode of spread in developed countries, reflecting infection among i.v. drug abusers who share unsterilized needles.[3] As of December 31, 1992, a cumulative total of 76,696 cases of AIDS has been reported from Europe, with 34% of them injected drug users. Italy and Spain are the two European countries with the highest prevalence of AIDS transmitted by i.v. drug abuse, showing 65.7% and 64.2% respectively.[4] A number of studies have attempted to quantify the likelihood of transmission in various circumstances. The Centers for Disease Control have estimated that the probability of transmission from a single anal receptive sexual contact with an infected partner is 1 in 100 to 1 in 500, while a single needle stick is thought to transmit the virus 1 in 1000 times.[5] However, there is considerable individual-to-individual variability resulting from the interplay of route, cofactors, and viral factors that affect the probability of any single infection taking place. Among the cofactors intervening in

0-8493-7842-7/94/$0.00+$.50

© 1994 by CRC Press, Inc.

these processes, malnutrition may play an important role. Thus, both nutritional deficiencies and their causes have a great relevance to these infectious conditions. According to Mgone et al.,[6] the prevalence of HIV infection in children may depend on the extent of the malnutrition status, being more severe in marasmus than in kwashiorkor.

In drug addicts, not only life-style, but anorexic effects of the drug lead to an extremely compromised nutritional status. Therefore, Altés et al.[7] found a prevalence of protein-energy malnutrition in drug addicts who had been receiving i.v. heroin until a few hours prior to their hospital admission for detoxification. These nutritional deficiencies can impair immunity and, thus, influence susceptibility to infectious agents, including HIV.[8]

Although therapy in i.v. drug addicts has a clear, positive benefit, new problems are continuously arising. A challenge to be met today is to recruit either infected or non-infected patients in order to assess their clinical nutritional status with a certain viability, so that they can afterwards be submitted to either a rehabilitation or nutritional advice program. Even when they have voluntarily decided to withdraw from drugs, the challenges will be present, unless they join up with an institutionalized center, following established rules. Therefore, most of the studies carried out on nutritional status of drug addicts have been performed during detoxification periods.

This chapter reviews the deleterious effects of drug abuse on nutrition and the immune system (both in humans and in experimental models); the role played by the immunological mediators in hypermetabolism; and its incidence on the nutritional status of HIV patients. The most recent reports related to the nutritional assessment of HIV-positive drug addicts are discussed in order to clarify the interrelationships between nutritional and immunological status in drug addicts.

NUTRITIONAL AND IMMUNOLOGICAL EFFECTS OF ABUSE SUBSTANCES

It is well documented that the use of addictive drugs, such as heroin, cocaine, marijuana, etc, affects food and liquid intake behavior, and taste preference.[9] These alterations may often lead to severe weight loss, which produces body deterioration and contributes to the development of malignant or infectious complications.

Male opiate addicts have been found to present a peculiar preference for sweets, tending to replace foods that are rich in fat and proteins with foods rich in sucrose and relatively poor in vitamins and minerals. However, the caloric intake of drug addicts was similar to that of non-addicts. These food habits gradually impair the nutritional status of drug addicts.[10] This study provides quantitative evaluation, in terms of nutrient intake, of the typical craving described by opiate addicts.

Cocaine use has increased over the past decade, raising concerns about its effect on host defenses. Cocaine injected into mice for 10 days has been shown to reduce the number of T cells, increasing B cells, and having no effect on macrophages in the spleen.[11] According to these authors, cocaine and malnutrition are involved in HIV infection. When the authors manipulate the mice immune suppression, the development of AIDS could be accelerated by both drug and nutritional deficit.

Watson et al.[12] investigated the combined effects of protein undernutrition and cocaine injection during retrovirus infection on the number of T cell subsets in mice. C57BL/6 female mice were fed a 4% casein diet, together with a progressively increasing daily cocaine i.p. injection for 11 weeks, and retroviral infection with LP-BM5. Cocaine administration reduced body and spleen weights, and partially prevented the lymphoid organs enlargement due to lymphoid cell proliferation induced by the murine retrovirus infection. As to the spleen, cocaine-injected mice showed an increase in the percentage of CD8+ cells and B cells and retrovirus-infected mice presented a decrease in the percentage of Thy 1.2+ and CD8+ cells and an increase in the percentage of CD4+ and B cells. Changes were more dramatic in the thymus gland. Retrovirus and cocaine "in vivo" altered synthesis of tumor necrosis factor and interferon-gamma by spleen cells stimulated with mitogen "in vitro". Thus, the principal effect of cocaine in retrovirally infected mice was to significantly reduce the number of cells with activation markers and cell proliferation.

Administration of opiates and opioid peptides has been reported to affect a wide range of immunological changes.[13] For example, a subcutaneous implant of morphine, commonly used, produces sustained atrophy of the spleen and thymus.[14] This atrophy was manifest within 24 h after implantation and was accompanied by a decrease in lymphocyte content in spleen and thymus,[15] inhibition of mitogen-stimulated T and B cell response,[16] and altered antigen-specific antibody production.[17] Elevation of intracellular calcium ions has been proposed to be one of the early events in a series of intracellular processes which culminate in lymphocyte activation and proliferation.[18] Sei et al.[19] have suggested that the morphine-induced inhibition of calcium ions mobilization in immune cells may be one of the early events mediating opiate-induced immune suppression.

López et al.[20] have recently developed an experimental model which resembles human drug addiction, in order to study the effect of chronic drug (cocaine or morphine) administration on the immune system of mice that received either a 20% or 4% casein diet. Both drugs were administered in increasing daily doses. Cocaine reduced body and spleen weights, particularly in the low protein diet group. Cocaine-injected mice showed a decrease in the percentage of CD4+, CD8+ and Mac-1 cells and an increase in B cells in the spleen of well-nourished mice. Morphine-treated mice showed similar results to those observed in cocaine-treated mice. These results suggest that cocaine or morphine injection can alter the percentage of cells that express a defined phenotype, regardless of the nutritional status of the subject.

Varela et al.[21] and Marcos et al.[22] showed that in male heroin addicts under detoxification (3–6 months), in spite of weight recovery after this period, the sequelae from the malnutrition status secondary to heroin abuse persists during this detoxification period.

According to the literature consulted, all the authors confirm that abuse of drugs not only can induce an anorexigen effect, but can also provoke an immune depletion. These two mechanisms come together in the drug abuser, leading to an interactive process among drug abuse, malnutrition, immunosuppression and infection susceptibility (e.g.: HIV) (Figure 1). All these factors are capable of creating a favorable atmosphere for triggering other clinical complications. The following step to be considered should be to what extent infection processes, especially HIV, are capable of producing malnutrition situations. This issue will be dealt with in the next section.

MALNUTRITION IN HIV INFECTION

Evolution of AIDS is often characterized by a major weight loss, which eventually leads to extreme caquexia. The origin and the causes of this malnutrition in AIDS are unclear. As in other chronic diseases, the development of malnutrition might contribute to a clinical dysfunction of the immune reaction and be an aggravating factor for the progression of diseases.[23] In the course of AIDS, part of the loss in body weight can be attributed to acute catabolic states related to fever or to evolving acute infections.

Malnutrition may also precede the diagnosis of AIDS; thus, reduced food intake as well as increased digestive loss of nutrients and energy expenditure are among possible mechanisms for progressive wasting.[24] The body weight loss in clinically stable but malnourished HIV-infected patients has been shown to be, at least, partly mediated by an increased resting energy expenditure (REE), which gives rise to a hypermetabolic state.[25] This hypermetabolic status also occurs in well-nourished HIV-infected subjects. To this effect, Hommes et al.[26] showed increased REE and rates of fat oxidation in men infected with HIV. Although the men had been antiHIV antibody positive for one year, circulating CD4+ T cell levels were normal and the patients were clinically asymptomatic, well-nourished, and classified as CDC group II or III. Albeit hypermetabolism can be mediated by elevated concentrations of thyroid hormones, catecholamines, or cortisol,[27] Hommes et al.[26] indicate that no patients showed higher concentrations of catabolic hormones, thereby suggesting that HIV infection causes metabolic alterations through another mechanism. Thus, cytokines, like tumor necrosis factor (TNF), interleukin 1 (IL-1), and IL-6 might be involved in this alternative mechanism. TNF and IL-1 appear to have metabolic effects related to tissue wasting;[28] IL-6 induces the hepatic acute phase response to inflammation or tissue injury.[29] Hommes et al.[26] affirm that, although unlikely, the higher

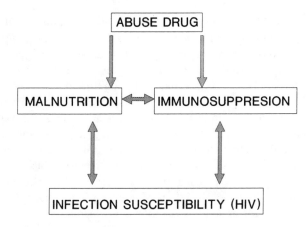

Figure 1   Interrelationships among abuse drug, malnutrition, immunosuppression, and infection susceptibility in HIV-positive drug addicts.

concentration of IL-6 found in the patients, possibly in conjunction with other cytokines, was involved in the pathophysiological mechanism responsible for the hypermetabolism in HIV-infected patients.

Therefore, the risk of malnutrition threatens HIV-infected patients from the onset of the illness. Thus, the nutritional evaluation of these patients is of great importance, together with nutritional programs leading to the best possible nutritional status in order to delay AIDS development.

Regarding the subjects in this chapter, HIV-positive drug addicts, as we have previously explained, the effects of both drug and HIV infection impinge on the patients. In spite of the fact that the nutritional evaluation of these subjects is quite laborious, some research has been carried out.

NUTRITIONAL STATUS ASSESSMENT IN HIV-POSITIVE DRUG ADDICTS

The information related to nutritional status in HIV-infected patients ranging from asymptomatic to developed AIDS has progressively increased over the past decade. However, there is scarce literature about HIV-infected drug abusers. Varela et al.[21] assessed the nutritional status of 36 male heroin addicts, with a previous addiction time of 5–15 years, non-infected and HIV-infected, undergoing a period of detoxification (over 3 months). The nutritional evaluation was carried out through immune function evaluation, in order to find out which of the following causes — HIV infection, heroin abuse or nutritional status — was the main one affecting the immune system. Height, weight and serum albumin concentration were measured, and immune function was tested using a delayed hypersensitivity skin test. No significant differences were found in anthropometric measurements between both groups, but an important anthropometric improvement (weight gain between 8 and 13 kg) was seen in the patients following the detoxification period (Table 1). Serum albumin, often used as a classical index of malnutrition, remained within the normal values in every patient. The overall response to the skin test was found to be depressed in both groups and no significant differences were shown between them (Table 2). These results may suggest that in spite of weight recovery after the detoxification period in the drug addicts evaluated, the subclinical symptoms of malnutrition secondary to heroin abuse persist. Nevertheless, the depleted nutritional status, assessed by the parameters cited above, might remain unmodified by HIV infection in the drug addicts tested. In a parallel work, under similar conditions, in which a more profound study of the immune system was performed, although no changes were found in leukocyte and lymphocyte counts between non-infected and HIV-infected drug addicts, alterations in lymphocyte subsets (CD4+ and natural killer cells were lower and CD8+ and B lymphocyte were higher than in control subjects) have been shown as a consequence of the drug consumption. Furthermore, a significant lower B lymphocyte rate and CD4/CD8 ratio were found in HIV-positive drug addicts in comparison with the HIV negative group (Table 3). These data suggest that, despite the anthropometric improvement shown following the detoxification period, a depleted immune status was observed in the studied drug addicts, mainly due to the subclinical malnutrition state triggered by drug abuse, but also as a consequence of HIV infection.[30]

In the revised literature, most of the reports referred to in AIDS and HIV infection have been determined in males, maybe due to the higher incidence (82.5% AIDS male patients according to Spanish data) in comparison with women. These data are very similar to those found in the Spanish drug addict population (81% males vs 19% females). On the other hand, a relevant issue to be considered is the higher risk for babies from HIV-infected mothers, since at least in Spain 78% of seropositive infants come from HIV-seropositive drug addict mothers.[31] Therefore, a study on a population of drug addicts (men and women, HIV-positive and HIV-negative) was carried out.[22] The nutritional status of these patients, who were undergoing one month of detoxification, was evaluated by dietary intake, anthropometric and immunological parameters (lymphocyte subsets, delayed hypersensitivity skin test and immunoglobulin serum rate). An important depletion of their immune function was shown, which might mean a subclinical status of malnutrition in all the drug addicts tested, being more noticeable in HIV-infected women. Nevertheless, the anthropometric measurements improved after the detoxification period in all patients. In spite of the fact that nutrient intake was usually higher in positive HIV groups (males and females), their weight gain was lower than in non-infected HIV ones. All females showed a lower iron intake than the RDA

TABLE 1.
Effect of HIV Infection on Anthropometric Parameters in Heroin Addicts Keeping Detoxification Therapy (3 Months)
(Means ± Sem)

	HIV (–) (n = 20)	HIV (+) (n = 16)
Age (years)	25.25 ± 0.81	25.40 ± 1.02
Actual weight (kg)	66.32 ± 1.67	67.38 ± 1.46
Previous weight (kg)	56.52 ± 1.87	55.08 ± 1.53
Height (cm)	171.66 ± 1.87	172.75 ± 0.91
Actual BMI (kg/m^2)	22.52 ± 0.46	22.69 ± 0.45
Previous BMI (kg/m^2)	19.17 ± 0.48	18.68 ± 0.42
Weight recovered (kg)	9.79 ± 1.43	12.45 ± 1.02
Weight recovered (%)	19.46 ± 3.83	23.06 ± 2.23

No significant differences between columns.

From Varela, P. et al., *Eur. J. Nutr.*, 44, 415, 1990. With permission.

TABLE 2.
Effect of HIV Infection on the Response to Delayed Cutaneous Hypersensitivity Test in Heroin Addicts Undergoing Detoxification Therapy (3 Months)

	Score	HIV (–) n = 20	HIV (+) n = 16
Anergy (%)	0	10	14
Relative anergy (%)	1 +ve response	5	36
Hypoergy (%)	< 10 mm	60	50
Low response (%)	10–20 mm	25	0
Normal response (%)	>20 mm	0	0
N^o of positive response to 7 antigens (means ± sem)		2.55 ± 0.34	1.73 ± 0.32
Score (mm, means ± sem)		6.38 ± 0.84	4.86 ± 0.97

No significant differences between columns. The distribution of scores is significantly different between both groups (Wilcoxon 2-sample test: $P = 0.013$).

From Varela, P. et al., *Eur. J. Nutr.*, 44, 415, 1990. With permission.

(recommended daily dietary allowance), while zinc and magnesium intakes were deficient in all patients studied. Vitamin C intake did not reach RDA in HIV-infected males. Harakeh et al.,[32] testing the action of ascorbic acid on HIV activity in a chronically HIV-infected T lymphocytic cell line, found that the presence of nontoxic concentrations of ascorbic acid in the cell culture medium reduced the level of extracellular reverse transcriptase activity by 99% and the

TABLE 3.
Lymphocyte Subpopulations Values in Control, HIV (–) and
HIV (+) Drug Addicts Under Detoxification Therapy
(3–6 Months) (Means ± sd).

	Control n = 13	HIV (–) n = 25	HIV (+) n = 15	Anova
CD2 ($\times 10^9$/L)	1.72 ± 0.97	1.80 ± 0.68	1.73 ± 0.82	NS
CD2 (1)	0.74 ± 0.07	0.63 ± 0.17*	0.71 ± 0.15	NS
CD4 ($\times 10^9$/L)	1.14 ± 0.03	1.06 ± 0.30	0.73 ± 0.38#	P <0.001
CD4 (1)	0.50 ± 0.01	0.39 ± 0.06*	0.28 ± 0.09#	P <0.001
CD8 ($\times 10^9$/L)	0.60 ± 0.07	0.80 ± 0.30*	1.06 ± 0.48	P <0.001
CD8 (1)	0.27 ± 0.03	0.29 ± 0.08	0.44 ± 0.09	P <0.001
CD4/CD8	1.90 ± 0.22	1.45 ± 0.51*	0.70 ± 0.41#	P <0.001
CD19 ($\times 10^9$/L)	0.19 ± 0.02	0.26 ± 0.15*	0.18 ± 0.13	NS
CD19 (1)	0.08 ± 0.01	0.09 ± 0.04	0.06 ± 0.04	NS
CD57 ($\times 10^9$/L)	0.26 ± 0.02	0.08 ± 0.06	0.06 ± 0.04	P <0.001
CD57 (1)	0.11 ± 0.01	0.03 ± 0.02*	0.03 ± 0.02	P <0.001

* Significantly different compared to control (p <0.05). # Significantly different compared to the HIV (–) group (p <0.05). NS: not significant.

From Marcos, A. et al., *Immunol. Infect. Dis.,* in press. With permission.

expression of p24 antigen (a measure of HIV protein) by 90%. Moreover, the presence of ascorbic acid caused a time- and dose-dependent decrease in HIV activity in acutely infected CD4+ T lymphocytes. Bodgen et al.[33] observed serum vitamin and mineral concentrations in 30 subjects with AIDS, AIDS-related complex or asymptomatic infection with HIV. They found that one or more abnormally low concentrations of the plasma micronutrients studied are likely to be present in the majority of HIV seropositive patients. According to McCorkindale et al.,[34] with the exception of a decrease in vitamin B_6, zinc, and total energy intake, food records closely matched the RDA in 19 patients infected with HIV for up to 16 months.

CONCLUSIONS

It is important for healthcare professionals caring for HIV-infected drug addicts to bear in mind the relationships among nutrition, drug abuse, HIV infection and the immunological status. Therefore, the nutritional status and clinical history should be assessed in each HIV-positive drug addict as early as possible. Each patient should be given nutritional education and therapy in order to improve their nutritional status, thereby delaying disease progression, helping them to get out of the drug habit and significantly contributing to an improved quality of life.

REFERENCES

1. WHO. (1992) *Weekly Epidemiological Record.* 67:201–204.
2. Friedland GH, Harris C, Butkus-Small C, Shine D, Moll B, Darrow W, and Klein RS. (1985) Intravenous drug abusers and the acquired immunodeficiency syndrome (AIDS). *Arch. Intern. Med.* 145:1413–1417.
3. Blattner WA. (1991) HIV epidemiology: past, present, and future. *FASEB J.* 5:2340–2348.
4. European Center for Epidemiological Monitoring of AIDS (1992) AIDS Surveillance in Europe. Quarterly Report nº 35.
5. Control Disease Center. (1987) Update: human immunodeficiency virus infection in health-care workers exposed to blood of infected patients. *Morbid Mortal Wkly Rep.* 36:285–289.
6. Mgone CS, Mhalu FS, Shao JF, Britton S, Sandstrom A, Bredberg-Raden U, and Biberfeld G. (1991) Prevalence of HIV-1 infection and symptomatology of AIDS in severely malnourished children in Dar Es Salaam, Tanzania. *J Acquir Immune Defic Syndr.* 4:910–913.
7. Altés J, Dolz C, Obrador A, and Forteza-Rei J. (1988) Prevalence of protein-energy malnutrition in heroin addicts hospitalized for detoxification. *J Clin Nutr Gastroenterol.* 3:55–58.
8. Moseson M, Zeleniuch-Jacquotte A, Belsito DV, Shore RE, Marmor M, and Pasternack B. (1989) The potential role of nutritional factors in the induction of immunologic abnormalities in HIV-positive homosexual men. *J Acquir Immune Defic Syndr.* 2:235–247.
9. Mohs ME, Watson RR, and Leonard-Green T. (1990) Nutritional effects of marijuana, heroin, cocaine, and nicotine. *J Am Diet Assoc.* 90:1261–1267.
10. Morabia A, Fabre J, Chee E, Zeger S, Orsat E, and Robert A. (1989) Diet and opiate addiction: a quantitative assessment of the diet of non-institutionalized opiate addicts. *B J Addict.* 84: 173–180.
11. Pillai RM and Watson RR. (1990) In vitro immunotoxicology and immunopharmacology: Studies on drug abuse. *Toxicol Letters.* 53:269–283.
12. Watson RR, Chen G, and López MC. (1992) Effects of protein undernutrition and cocaine injection on T-lymphocyte numbers in retrovirally infected mice. In: *Nutr Immunology.* Chandra RK, ed. pp 269–281. ARTS Biomedical Publishers and Distributors. St. John's, Newfoundland, Canada.
13. Sibinga NES and Goldstein A. (1988) Opioid peptides and opioid receptors in cells of the immune system. *Annu Rev Immunol.* 6:219–249.
14. Arora PK, Fride E, Petitto J, Waggie K, and Skolnick P. (1990) Morphine-induced immune alteration in vivo. *Cell Immunol.* 126:343–353.
15. Sei Y, Yoshimoto K, McIntyre T, Skolnick P, and Arora PK. (1991a) Morphine-induced thymic hypoplasia is glucocorticoid-dependent. *J Immunol.* 146:194–198.
16. Bryant HU, Berton EW, and Holaday JW. (1988) Morphine pellet-induced immunomodulation in mice: temporal relationships. *J Pharmacol Exp Ther.* 245:913–920.
17. Weber RJ, Ikejiri B, Rice KC, Pert A, and Hagan AA. (1987) Opiate receptor mediated regulation of the immune response in vivo. *Natl Inst Drug Abuse Res Monogr Ser.* 76:341–348.

18. Linch DC, Wallace DL, and O'Flynn K. (1987) Signal transduction in human T lymphocytes. *Immunol Rev.* 95:137–159.

19. Sei Y, McIntyre T, Fride E, Yoshimoto K, Skolnick P, and Arora PK. (1991b) Inhibition of calcium mobilization is an early event in opiate-induced immuno-suppression. *FASEB J.* 5:2194–2199.

20. López MC, Huang DS, Waltz B, Chen GJ, and Watson RR. (1991) Splenocyte subsets in normal and protein malnourished mice after long-term exposure to cocaine or morphine. *Life Sci.* 49:1253–1262.

21. Varela P, Marcos A, Ripoll S, Requejo A, Herrera P, and Casas A. (1990). Nutritional status assessment of HIV-positive drug addicts. *Eur J Clin Nutr.* 44:415–418.

22. Marcos A, Varela P, Requejo AM, Casco A, Santacruz I, Montes F, and Arce MM. (1992) Los parámetros inmunológicos como evaluadores de malnutrición subclínica en drogodependientes. *An Real Acad Farm.* 58:415–427.

23. Kotler DP, Tierney AR, Wang J, and Pierson RE Jr. (1989) Magnitude of body cell mass depletion and the timing of death from wasting in AIDS. *Am J Clin Nutr.* 50:444–447.

24. Kotler DP, Wang J, and Pierson R. (1985) Studies of body composition in patients with the acquired immunodeficiency syndrome. *Am J Clin Nutr.* 42:1255–1265.

25. Melchior JC, Salmon D, Rigaud D, Leport C, Bouvet E, Detruchis P., Vildé JL, Vachon F, Coulaud JP, and Apfelbaum M. (1991) Resting energy expenditure is increased in stable, malnourished HIV-infected patients. *Am J Clin Nutr* 53:437–441.

26. Hommes MJT, Romijn JA, Endert E, and Sauerwein HP. (1991) Resting energy expenditure and substrate oxidation in human immunodeficiency virus (HIV)-infected asymptomatic men: HIV affects host metabolism in the early asymptomatic stage. *Am J Clin Nutr.* 54:311–315.

27. Bessey PQ, Watters JM, Aoki TT, and Wilmore DW. (1984) Combined hormonal infusion simulates the metabolic response to injury. *Ann Surg.* 200:264–281.

28. Evans RD, Argilés JM, and Williamson DH. (1989) Metabolic effects of tumour necrosis factor-alpha (cachectin) and interleukin-1. *Clin Sci.* 77:357–364.

29. Kishimoto T. (1989) The biology of interleukin-6. *Blood* 74:1–10.

30. Marcos A, Varela P, and Santacruz I. (1993) Nutritional status in non-infected and human immunodeficiency virus infected asymptomatic drug addicts. Immunological assessment. *Immunol Infect Dis* (in press).

31. Ministerio de Sanidad y Consumo. (1992) Vigilancia del SIDA en España. Informe Trimestral n° 4/92. Madrid. España.

32. Harakeh S, Jariwalla RJ, and Pauling L. (1990) Suppression of human immuno-deficiency virus replication by ascorbate in chronically and acute infected cells. *Proc Natl Acad Sci USA.* 87:1745–1749.

33. Bodgen JD, Baker H, Frank O, Perez G, Kemp F, Bruening K, and Louria D. (1990) Micronutrient status and human immunodeficiency virus (HIV) infection. *Ann N Y Acad Sci.* 587:189–195.

34. McCorkindale C, Dybevik K, Coulston, AM, and Sucher KP. (1990) Nutritional status of HIV-infected patients during the early disease stages. *J Am Diet Assoc.* 90:1236–1241.

Chapter

5

Energy Metabolism in HIV-Infected Patients

Jean-Claude Melchior

INTRODUCTION

Malnutrition is one of the major complications of Acquired Immunodeficiency Syndrome (AIDS). Weight loss is often one of the earliest clinical manifestations of the illness and may ultimately reach cachexia. It is a major cause of morbidity, independent of immune deficiency, and contributes to the diminished life quality of the patients. Protein-energy malnutrition (PEM) is the first cause of immunodeficiency in underdeveloped countries. Hughes et al[1] have shown that PEM increases the incidence of Pneumocystis Carinii pulmonary infection in laboratory rats as well as in infants suffering from immunodeficiency related cancer. However, in AIDS malnutrition is not evidently correlated with immunodeficiency as weight loss is not related to CD4 cells count (Figure 1). The development of malnutrition may further contribute to the clinical dysfunction of the immune reaction and be an aggravating factor for the progression of the disease.[2-4] Moreover, severe malnutrition may shorten life expectancy insofar as death has been specifically related to the loss of body cell mass.[5] To maintain weight, total energy expenditure must equal caloric intake. If caloric intake exceeds total energy expenditure, the excess calories are stored as either cellular mass or fat. On the other hand, if total energy expenditure exceeds the calories absorbed, an energy deficit results that leads to the breakdown of protein or fat for production of energy.

0-8493-7842-7/94/$0.00+$.50
© 1994 by CRC Press, Inc.

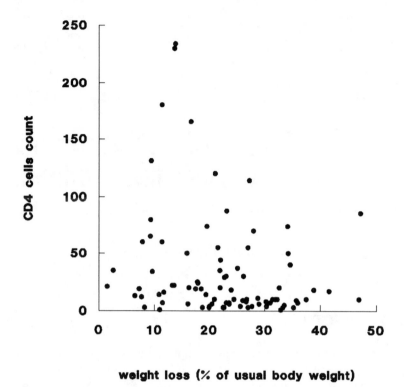

Figure 1.   Relation between body weight loss and immunity in 85 HIV infected patients (personal data). Immunity is represented by the CD4 cells count. Weight loss is expressed as percentage of usual body weight.

METABOLIC ADAPTATION TO DEFICIT OF ENERGY INTAKE

In animals as well as in man, energy deficit due to semi-starvation is met by an integrated adaptive response. The hormonal response leads to sparing the protein cell mass and to use of the fat mass for energy. The adaptive response during long-term fasting in emperor penguins gives some insight into the integrated response to fasting or semi-starvation (Figure 2). After a short period of adaptation, which is similar to the human response, protein utilization decreases, and most of the energy used derives from lipids.[6] This fact is a major determinant in the ability to tolerate longer fasting periods. A third and ultimate phase, when half of initial body cell mass is lost, leads to the death of the animal. This period is characterized by increased protein oxydation just before death. In the normoponderal man, adipose tissue is an important energy reserve which can contribute to the preservation of cell mass in the case of a deficient energy balance (Figure 3). During starvation in man, death occurs for a body weight loss

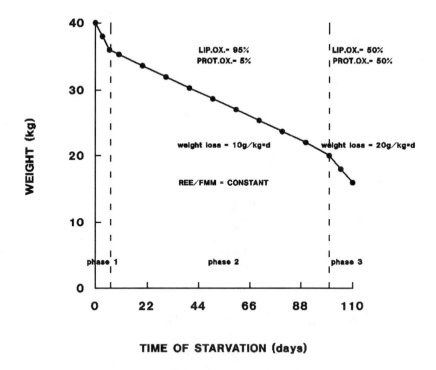

Figure 2. Adaptive response during fasting in emperor penguins. (Adapted from Le Maho, Y., *Am. J. Physiol.,* 254, 61, 1988.) Lip. Ox. is the percentage of lipids included in the total daily energy expenditure. Prot. Ox. is the percentage of Proteins included in the total daily energy expenditure. Weight loss is expressed in g/kg·d.

amounting to about 45% of ideal body weight.[7-9] On the other hand, during the flow phase of sepsis or trauma, the hormonal response fails to spare the protein mass. The wasting, in that case, bears predominantly on muscular protein stores as it does in the protein malnutrition of children, i.e. kwashiorkor.[10-12]

BODY COMPOSITION AND SURVIVAL IN AIDS

The study of body composition is of particular interest to the evaluation of malnutrition in HIV-infected patients since Kotler observed that survival is correlated with the magnitude of body cell mass depletion.[5] In this study, at the time of death, both extrapolated and observed values for body cell mass were 54% of normal. However, malnutrition-induced changes affect both the fat and the fat free mass (FFM). Several methods are available to determine body composition. Anthropometric determination is the most largely used method, but it fails to estimate the changes of either extra cellular fluid or protein stores.

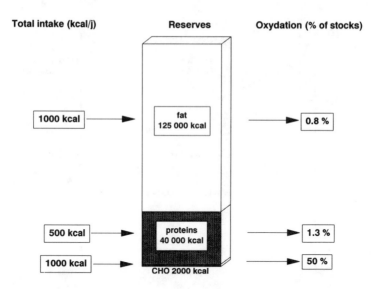

Figure 3.   Energy reserves in Man and relative contribution in total daily energy balance.

Altogether, the determination of body-fat and FFM are comparable with the different methods used,[13] but bioelectric impedance (BIA) appears to be one of the more accurate and practical methods. It allows determination of total water, and even, using two frequencies, of extra- and intra-cellular water.[14] It is both economical and non-invasive, and can be used at the bedside. For example, Figure 4 presents the body composition of 119 HIV-infected patients compared with a control group (personal data). The correlation between FFM measured by BIA and anthropometric determination was 0.96. However BIA, measured by the Boulier's method using two frequencies, evidenced a higher extra-cellular fluid and a relative decrease of intra-cellular fluid and proteins in HIV-infected patients. These results were confirmed by other publications using several methods for determination of body composition.[15,16] Further, the use of tetrapolar body impedance analysis shows significant early malnutrition in HIV-infected patients even in the absence of body weight loss.[17]

ENERGY BALANCE AND HIV

The development of impaired energy balance in AIDS is multifactorial. Food intake may diminish for a variety of reasons.[18,19] Nutrient malabsorption which often occurs in the early stage of the illness is well documented.[20,21] Increased energy expenditure is also one of the possible mechanisms for progressive wasting. In fact, many patients suffer from a progressive weight loss during periods presumably free of secondary infections, in the course of what has been called "the wasting syndrome".[22] Total energy expenditure

BODY COMPOSITION (119 PATIENTS HIV+)
BIOELECTRIC-IMPEDANCE (1mHz et 5kHz)

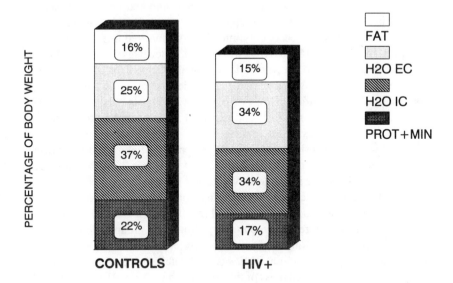

Figure 4. Body composition of 119 HIV infected patients compared with a control group (personal data). Fat is the fat mass. H_2O EC is the extracellular fluid. H_2O IC is the intracellular fluid and Prot + Min is the sum of proteins and minerals. Measures are performed according to Boulier's method.[14]

comprises resting energy expenditure (REE) (which accounts for about 60–75% of total energy expenditure), plus dietary thermogenesis (about 10–15% of total) plus energy related to activity (about 10–30% of total). The main determinant of REE is the FFM which must be taken into account when studying the variations of REE in pathology.[23,24] In acute and chronic starvation, REE decreases slowly, but after a short period of adaptation, the decrease is mainly dependent on the loss of body cell mass.[25] On the contrary, during the flow phase of sepsis and/or trauma, the REE is usually increased[10] and the nitrogen balance severely negative. Both may persist despite aggressive hyperalimentation.[26] This contrasts with weight loss associated with some forms of cancer in which there is no hypermetabolism, and in which nitrogen balance can be restored by hyperalimentation. In most cases of cancer, trauma or infections, metabolic disturbances range between these extremes. AIDS is characterized by primary infection with HIV followed by secondary infections and cancers. In order to understand the wasting process in AIDS, one must consider both the metabolic disturbances due to HIV infection itself and those due to secondary illnesses. In six patients, Kotler's group did not find an elevated REE compared with theoretical values obtained by use of Harris and Benedict's equation.[27] Despite this result, it is now not doubted that REE is elevated in the course of HIV infection in view of results reported by our group and others, even in asymptomatic patients with normal CD4 lymphocytes counts.[28,29,30,31] We measured the REE of three hundred and thirty patients divided in two groups. The first group was composed of two hundred and sixty patients free of evolutive secondary infection. All suffered from AIDS-related complex (ARC; n = 44) or AIDS (n = 216). The diagnosis of AIDS was established according to the criteria defined by the Centers for Disease Control and Prevention.[32] Nutritional deficit was evidenced by a substantial weight loss, defined as a history of loss of over 10% of customary weight.[22] The relation between measured REE and expected value according to Harris and Benedict's equation was mediocre (r = 0.50). The correlation between REE and fat free mass in this group of patients is presented in Figure 5. As in normal subjects, FFM remains the first determinant of REE (r = 0.79), but in respect of FFM, REE is higher in stable HIV-infected patients than in controls. REE measured in HIV-infected patients suffering from a secondary evolutive infection (Table 1) remains dependent on FFM (r = 0.70), but at a higher level than in stable patients (Figure 6) as reported by Grunfeld's group and our own.[19,31] In a large group of HIV-infected patients considered to be in a stable state, i.e. devoid of developing infection, REE was moderately but clearly higher than in the control group. The weakness of the correlation between REE and expected values confirms the results of other studies in cancer patients and in critically ill patients.[10] Thus, when REE is to be compared among different groups of subjects, it must be referred to as an index of body size, i.e. FFM. This can be accurately measured with a BIA analyzer[14] or calculated from anthropometric measurements.[33]

ETIOLOGY OF HYPERMETABOLISM IN HIV-INFECTED PATIENTS

The origin of hypermetabolism remains unclear. It could be argued that, in clinically stable patients, it might be related to subclinical opportunistic infections. But, in our studies as in that of Grunfeld et al,[19,31] patients were free of secondary infection for over two weeks before and after REE measurements. Moreover, Hommes et al reported a similar increase in REE in asymptomatic HIV-infected patients.[30] Hypermetabolism can be mediated by elevated concentrations of thyroid hormones, catecholamines or cortisol.[11,28] The only reported concentrations of hormones in HIV patients were a normal serum triiodothyronine[34] and a low concentration of norepinephrine.[28] Such results cannot explain an increased REE. The possibility of an unknown mechanism, independent of elevated catabolic hormones, cannot be excluded.

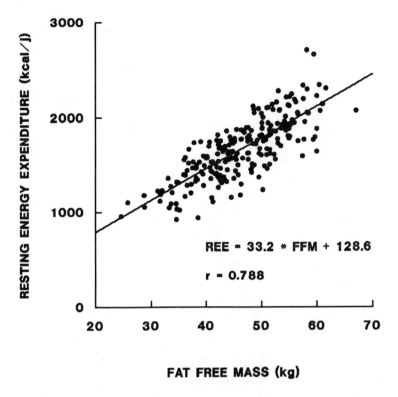

Figure 5. Regression between resting energy expenditure and fat free mass in human immuno-deficiency virus (HIV)-infected patients without secondary infection. n = 260 (personal data).

TABLE 1.
AIDS and Secondary Systemic Infections
(70 Patients)

Infections	Number of Case
Mycobacterium avium intracellulare	n = 23
Pneumocystis carinii pneumoniae	n = 5
Cytomegalovirus retinis and/or colitis	n = 11
Toxoplasmosis	n = 10
Other systemic bacteria	n = 10

Cytokines, such as tumor necrosis factor (TNF), interleukin 1 (IL-1), and IL-6 might be involved in this alternative mechanism.[35-40] TNF and IL-1 have metabolic effects related to tissue wasting,[41,42] while IL-6 induces the hepatic acute phase response to inflammation or tissue injury.[43] Conflicting reports on concentrations of TNF and IL-1 in HIV patients do not allow speculation

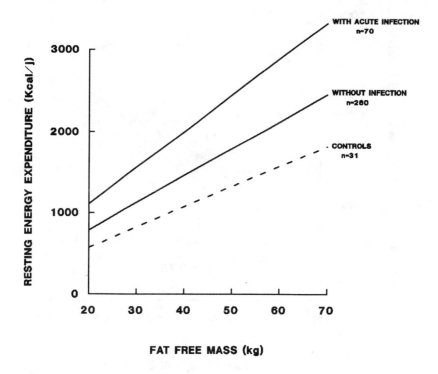

Figure 6. Comparison of the resting energy expenditure between the two groups of human immunodeficiency virus (HIV)-infected patients and control subjects expressed as a function of the fat free mass (personal data). Total HIV infected patients, n = 330.

on the exact potential role of these cytokines either in the wasting syndrome or in increased REE.[39,40] In fact, in a previous work, we could not provide evidence of an increase in plasma TNF concentrations in HIV patients with documented increases in REE.[44] Still, a role for TNF cannot be excluded in patients with evolving secondary infections.[45] In the absence of secondary infection, both synthesis and breakdown of protein seems to decrease in patients with AIDS.[46] This pattern does not resemble the increased protein turn over of overwhelming infections but instead is more like adaptation to starvation. On the other hand, certain secondary evolutive infections induce an even higher level of REE.

In massive infections, sepsis or burns, elevated REE is associated with accelerated protein catabolism despite a decrease in caloric intake. Preliminary data suggest that accelerated protein breakdown and negative nitrogen balance also occur in patients with AIDS when they suffer from secondary active infections.

CONSEQUENCES OF INCREASED REE ON NUTRITIONAL STATUS AND ENERGY BALANCE IN HIV-INFECTED PATIENTS

Despite a higher REE in stable HIV-infected patients, nitrogen balance is not deeply negative. Thus the patients could theoretically maintain a stable body weight especially if the level of food intake were preserved.[27] However, the decrease of protein synthesis may make it difficult to maintain lean muscle mass or regain it after rapid weight loss. Moreover, when the patients suffer from secondary infection, the higher REE associated with accelerated protein breakdown is also concomitant with a decreased food intake such that energy balance cannot be maintained, and weight loss could be accelerated above 5% in one month.[19]

STRATEGY AND PERSPECTIVE TO REVERSE WASTING

Since weight loss is more dramatic during secondary infection,[19] the search for diagnosis and treatment of the intercurrent infection is essential in all patients with AIDS who are wasting. The REE can be useful in the detection of secondary infection. In particular an elevated REE/FFM ratio is a good indicator for the presence of a secondary evolutive infection (Table 2). Successful treatment of systemic secondary infections results in gains in both weight and body cell mass.[47] There is some evidence that antiretroviral therapy of HIV infection itself might lead to substantial weight gain.[48] However, the effects of antiretroviral therapy on energy expenditure remain unknown. In the absence of specific treatment for secondary infections, hyperalimentation leads to increased body fat but fails to replete protein stores.[49] In stable patients with progressive wasting, the weight recovery is often difficult but seems possible using hyperalimentation.[50] The benefits possibly derived from anabolic therapies need to be explored. Anti-TNF or interleukin-1 receptor antagonists are the most likely candidates.[51] However, the targets and impact of the cytokine response to HIV infection remain largely unknown. Another anabolic approach would be the use of Megestrol acetate.[52,53] This may increase food intake and promote weight gain, but it is not yet clear whether this would represent increased muscle mass, water retention or fat deposit.[52]

Although growth hormone (GH) levels are similar in patients with AIDS and normal control, GH therapy has been proposed as an alternative approach of anabolic therapy,[54,55] but the effects of GH on HIV virus replication are

TABLE 2.
Discriminating Power of REE Toward Infection

	HIV Patients With Acute Infection	HIV Patients Without Acute Infection	
REE > 42.3 kcal·kg FFM-1·d-1	62	15	77
REE ≤ 42.3 kcal·kg FFM-1·d-1	8	245	253
Total	70	260	330

Note: Sensitivity = 89%; Specitivity = 94%; % of patients well classified = 93%.

unknown and the greatest caution is necessary. The moderately increased REE of stable patients justifies the exploration of hyperalimentation by means of enteral or total parenteral nutrition as an adjuvant until a specific treatment against the virus itself is available despite the difficulty involved with the practice of such treatments.[49,50,56]

REFERENCES

1. Hughes WT, Price RA, Sisko F, Havron WS, Kafatos AG, Schonland M, and Smythe PM. Protein-calorie malnutrition: a host determinant for pneumocystis carinii infection. *Am. J. Dis. Child.* 1974; 128: 44–52.
2. Chandra RK and Scrimshaw NS. Nutrition, immunity and infection: present knowledge and future directions. *Lancet* 1983; 1: 688–91.
3. Dossetor J, Whittle HC, and Greenwood BM. Persistent measles infection in malnourished children. *Br. Med. J.* 1977; 2: 1633–5.
4. Gray RH. Similarities between AIDS and PCM. *Am. J. Public Health* 1983; 73: 1332.
5. Kotler DP, Tierney AR, Wang J, and Pierson RE. Magnitude of body cell mass depletion and the timing of death from wasting in AIDS. *Am. J. Clin. Nutr.* 1989; 50:444–7.
6. Robin JP, Frain M, Sardet C, Grocolas R, and Le Maho Y. Protein and lipid utilization during long-term fasting in emperor penguins. *Am. J. Physiol.* 1988; 254: R61–R68.
7. Brozec J, Wells S, and Keys A. Medical aspects of semi starvation in Leningrad (siège 1941–1942). *Am. Rev. Soviet. Med.* 1946; 4: 70–86.
8. Fliederbaum J. Clinical aspects of Hunger disease in adults. In: *Hunger Disease: Studies by the Jewish Physicians in the Warsaw Ghetto.* Winick M, ed. Osnos M, trans. New York: John Wiley, 1979: 11–43.
9. Cahill GF. Starvation in man. *N. Engl. J. Med.* 1970; 282: 668–75.

10. Long CL, Schaffel N, Geiger JW, Schiller WR, and Blackemore WS. Metabolic response to injury and illness: estimation of energy and protein needs from indirect calorimetry and nitrogen balance. *J. Parent. Enteral. Nutr.* 1979; 3: 452–6.

11. Beisel WR. Metabolic response to infection. *Annu. Rev. Med.* 1975; 26: 9–20.

12. Rossouw JE and Pettifor JM. Protein energy malnutrition. In: *The Metabolic and Molecular Aspects of Acquired Diseases.* Cohen RD, Lewis B, Albuti KGMM, and Demman AM, eds., Baillere Tindall (London) 1990: 514–528.

13. Wang J, Kotler DP, Russel M, Burastero S, Mazariegos M, Thornton J, Dilmanian FA, and Pierson R. Body fat measurement in patients with acquired immunodeficiency syndrome: which method should be used? *Am. J. Clin. Nutr.* 1992; 56: 963–7.

14. Boulier A, Fricker J, Thomasset AL, and Apfelbaum M. Fat free mass estimation by the two-electrode impedance method. *Am. J. Clin Nutr.* 1990; 52(4): 581–5.

15. Kotler DP, Wang J, and Pierson R. Studies of body composition in patients with the acquired immunodeficiency syndrome. *Am. J. Clin. Nutr.* 1985; 42: 1255–65.

16. Melchior J-C. Aspects métaboliques de la dénutrition au cours du SIDA. *Nutrition Clinique et Métabolisme.* 1993; 7: 311–320.

17. Ott M, Lembcke B, Fischer H, Jäger R, Polat H, Geier H, Rech M, Staszeswki S, Helm EB, and Caspary WF. Early changes of body composition in human immunodeficiency virus-infected patients: tetrapolar body impedance analysis indicates significant malnutrition. *Am J Clin Nutr.* 1993; 57: 15–9

18. Dworkin BM, Wormser GP, Axelrod F, Pierre N, Schwartz E, Schwartz E, and Seaton T. Dietary Intake in patients with Acquired Immunodeficiency Syndrome (AIDS), patients with AIDS-related complex, and serologically positive human immunodeficiency virus patients: correlations with nutritional status. *J. Parent. Enteral. Nutr.* 1990; 14(6): 605–9.

19. Grunfeld C, Pang M, Shimizu L, Shigenada JK, Jensen P, and Feingold KR. Resting energy expenditure, caloric intake and weight charge in human immunodeficiency virus infection and the acquired immunodeficiency syndrome. *Am. J. Clin. Nutr.* 1992; 55: 455–60.

20. Martin Z, Ullrich R, Heise W, Bergs C, L'age M, and Riecken EO. Malabsorption is found in early stages of HIV infection and independent of secondary infections. In: *Abstracts of the Seventh International Conference on AIDS,* Florence, Italy, June 16–21, 1991. Vol. 1. Rome: Istituto Superiore di Sanità, 1991: 46. abstract.

21. Chlebowski RT, Grosvenor MB, Bernhard NH, Morales LS, and Bulcavage LM. Nutritional status, gastrointestinal dysfunction, and survival in patients with AIDS. *Am. J. Gastroenterol.* 1989; 84: 1288–93.

22. Grunfeld C and Kotler DP. The wasting syndrome and nutritional support in AIDS. *Sem. Gastrointest. Dis.* 1991; 2(1): 25–36.

23. Roza AM and Schizgal HM. The Harris Benedict equation reevaluated: resting energy requirement and the body cell mass. *Am. J. Clin. Nutr.* 1984; 40: 168–82.

24. Hoshino E., et al. Body composition and metabolic word missing, *Am. J. Physiol.* 1991; 260 (n°1): 27–36.

25. Melchior JC, Rigaud D, Rozen R, Malon D, and Apfelbaum M. Resting energy expenditure economy induced by decrease in lean body mass in anorexia nervosa. *Eur. J. Clin. Nutr.* 1989; 43: 793–9.
26. Burt ME, Stein TP, and Brennan MF. A controlled randomized trial evaluating the effects of enteral and parenteral nutrition on protein metabolism in cancer-bearing man. *J. Surg. Res.* 1983; 34: 303–14.
27. Kotler DP, Tierney AR, Brenner SK, Couture S, Wang J, and Pierson RN, Jr. Preservation of short-term energy balance in clinically stable patients with AIDS. *Am. J. Clin. Nutr.* 1990; 51: 7–13.
28. Hommes MJT, Romijn JA, Godfried MH, Eeftinc K, Schattenker JKM, Burman WA, Endert E, and Sauerwein HP. Increase resting energy expenditure in human immunodeficiency virus-infected men. *Metabolism* 1990; 39: 1186–90.
29. Melchior JC, Salmon D, Rigaud D, Leport C, Bouvet E, Detruchis P, Vilde JL, Vachon F, Coulaud JP, and Apfelbaum M. Resting energy expenditure is increased in stable, malnourished HIV-infected patients. *Am. J. Clin. Nutr.* 1991; 53: 437–41.
30. Hommes MJT, Romijn JA, Endert E, and Saverwein HP. Resting energy expenditure and substrate oxidation in human immunodeficiency virus (HIV)-infected asymptomatic men: HIV affects host metabolism in the early asymptomatic stage. *Am. J. Clin. Nutr.* 1991; 54: 311–5.
31. Melchior J-C., Raguin G., Boulier A., Bouvet E., Rigaud D., Matheron S., Casalino E., Vilde J-L., Vachon F., Coulaud J-P., and Apfelbaum M. Resting energy expenditure in HIV infected patients. Comparison between patients with and without secondary infections. *Am. J. Clin. Nutr.* 1993; 57: (in press).
32. Centers for disease control. Classification system for human T-lymphotropic virus type III/lymphodenopathy- associated virus infections. *MMWR* 1985; 35: 334–9.
33. Durnin JVGA and Womersley J. Body fat assessed from total body density and its estimation from skinfold thickness: measurements on 471 men and women aged from 16–72 years. *Br. J. Nutr.* 1974; 32: 77–9.
34. Lopresti JS, Fried JC, Spencer CA, and Nicoloff JT. Unique alterations of thyroid hormones indices in the acquired immunodeficiency syndrome (AIDS). *Ann. Intern. Med.* 1989; 110: 970–5.
35. Krauss RM, Grunfeld C, Doerrler WT, and Feingold KR. Tumor necrosis factor acutely increases plasma levels of very low density lipoproteins of normal size and composition. *Endocrinology* 1990; 127: 1016–21.
36. Semb H, Peterson J, Tavernier J, and Olivecrona T. Multiple effects of tumor necrosis factor on lipoprotein lipase in vivo. *J. Biol. Chem.* 1987; 262: 8390–4.
37. Grunfeld C, Gulli R, Moser AH, Gavin LA, and Feingold KR. Effect of tumor necrosis factor administration in vivo on lipoprotein lipase activity in various tissues of the rat. *J. Lipid. Res.* 1989; 30: 579–85.
38. Adi S, Pollock AS, Shigenaga JK, Moser AH, Feingold KR, and Grunfeld C. Role for monokines in the metabolic effects of endotoxin: interferon gamma restores responsiveness of C3H/HeJ mice in vivo. *J. Clin. Invest.* 1992; 89: 1603–9.

39. Feingold KR, Soued M, Serio MK, Moser AH, Fiers W, Dinarello CA, and Grunfeld C. Multiple cytokines stimulate hepatic lipid synthesis in vivo. *Endocrinology* 1989; 125: 267–74.

40. Grunfeld C, Adi S, Soued M, Moser AH, Fiers W, and Feingold KR. Search for mediators of the lipogenic effects of tumor necrosis factor: potential role of interleukin 6. *Cancer Res* 1990; 50: 4233–8.

41. Waage A, Espevick T, and Lamvick J. Detection of tumor necrosis factor-like cytotoxicity in serum from patients with septicaemia but not from untreated cancer patients. *Scand. J. Immunol.* 1986; 24: 739–43.

42. Gelin J, Andersson C, and Lundholm K. Role of endogenous tumor necrosis factor alpha and interleukin 1 for experimental tumor growth and the development of cancer cachexia. *Cancer Res.* 1991; 51(3): 880–5.

43. Feingold KR, Adi S, Staprams I, et al. Diet affects the mechanisms by which TNF stimulates hepatic triglyceride production. *Am. J. Physiol.* 1990; 259: E177–E184.

44. Salehian D, Salmon D, Nyangabo T, Melchior JC, Rigaud D, and Vilde JL. Tumor necrosis factor and resting energy expenditure during AIDS. *Lancet* 1992 (in press).

45. Lähdevirta J, Maury CPJ, Teppo A, and Repo H. Elevated levels of circulating cachectin/tumor necrosis factor in patients with acquired immunodeficiency syndrome. *Am. J. Med.* 1988; 85: 289–91.

46. Stein TP, Nutinsky C, Condoluci D, Schluter MD, and Leskiw MJ. Protein and energy substrate metabolism in AIDS patients. *Metabolism* 1990; 39: 876–81.

47. Kotler DP, Tierney AR, Atlilio D, Wang J, and Pierson RN. Body mass repletion during gangliovir treatment of cytomegalovirus infections in patients with the acquired immunodeficiency syndrome. *Arch. Intern. Med.* 1989; 149: 901–5.

48. Moore R, Hidalgo J, Sugland BW, and Chaisson RE. Zidovurine and the natural history of the acquired immunodeficiency syndrome. *N. Engl. J. Med.* 1991; 324: 1412–6.

49. Kotler DP, Tierney AR, Culpepper-Morgan JA, Wang J, and Pierson RN, Jr. Effects of home total parenteral nutrition upon body composition in patients with acquired immunodeficiency syndrome. *J. Parent. Enteral. Nutr.* 1990; 14: 454–8.

50. Kotler DP, Tierney AR, Ferraro R, et al. Enteral alimentation and repletion of body cell mass in malnourished patients with acquired immunodeficiency syndrome. *Am. J. Clin. Nutr.* 1991, 53: 149–154.

51. Sherry BA, Gelin J, Fong Y, et al. Anticachectin/tumor necrosis factor-alpha antibodies attenuate development of cachexia in tumor models. *FASEB J.* 1989; 3: 1956–62.

52. Von Roenn JH, Murphy RL, Weber KM, Williams LM, and Weitzman SA. Megestrol acetate for treatment of cachexia associated with human immunodeficiency virus (HIV) infection. *Ann. Intern. Med.* 1988; 109: 840–1.

53. Tierney AR, Cuff P, and Kotler DP. The effect of megestrol acetate (megace) on appetite, nutritional repletion and quality of life in AIDS cachexia. In: *Abstracts of the Seventh International Conference on AIDS,* Florence, Italy, June 16–21, 1991. Vol. 1. Rome: Istituto Superiore di Sanità, 1991: 231. abstract.

54. Krentz AJ, Koster FT, Crist FT, Finn K, Boyle PJ, and Schade DS. Beneficial
     anthropometric effects of human growth hormone in the treatment of AIDS. *Clin.
     Res.* 1991; 39: 220A. abstract.
55. Mulligan K, Grunfeld C, Hellerstein M, and Schambelan M. Growth hormone
     treatment of HIV-associated catabolism. *FASEB J.* 1992; 6: A1942. abstract.
56. Singer E, Rothkopf MM, Kvetan V, et al. Risks and benefits of home parenteral
     nutrition in acquired immunodeficiency syndrome. *J. Parent. Enteral. Nutr.*
     1991, 15: 75–79.

Chapter

6

Pathogenesis and Clinical Aspects of Diarrhea and Its Effect on Nutrition in Patients Infected with the Human Immunodeficiency Virus

Lawrence A. Cone

and

David R. Woodard

"Slim Disease"[1] is a highly descriptive African appellation for wasting in patients infected with the Human Immunodeficiency Virus (HIV). While clearly denoting loss of subcutaneous fat deposits and muscle mass, it does not imply a reason for the patient's willowy appearance. Yet it is apparent that the cause of malnutrition in persons with AIDS is multifactorial, consisting of decreased caloric intake, local and systemic infection, neoplasia and malabsorption. This review will restrict itself to the pathogenesis and clinical aspects of intestinal infestations associated with HIV infection and the accompanying malabsorption of nutrients and medications.

From the beginning of the AIDS pandemic more than a decade ago, it was observed that between 30% to 50% of patients in the USA and up to 90% of those in Africa and Haiti complain of chronic diarrhea.[2] Whereas an acute self-limited diarrhea occurs early after HIV infection (seroconversion diarrhea), more protracted low volume diarrhea or frank dysentery due to *Salmonella, Shigella, Campylobacter, Yersinia, Entamoeba* and Herpes simplex virus, which infect ileum, colon and rectosigmoid, occurs more commonly during the subtle evolution of HIV infection. With time, CD4+ cell counts are reduced

0-8493-7842-7/94/$0.00+$.50
© 1994 by CRC Press, Inc.

below 100/mm³, and a number of intestinal infections evolve which result in chronic high volume diarrhea and "slim" disease. These include *Cryptosporidium parvum,* microsporidia, mycobacteria, *Isospora belli* and cytomegalovirus.

SALMONELLA AND OTHER ENTEROPATHOGENIC INFECTIONS

The pathogenesis of Salmonella infections has become better understood in recent years. Intestinal mucosal immunity is quantitatively the largest [89%] and most discriminative aspect of the human immune system. It must be able to differentiate between nutrients and pathogens, allowing the former to pene- trate the host without initiating an immune response while destroying the latter. In an adult the synthetic rates in milligrams of immunoglobulin per kilogram of body weight per day are 66 for IgA, 34 for IgG, 7.9 for IgM, 0.4 for IgD and 0.02 for IgE. Serum IgA is predominantly derived from bone marrow; how- ever, the dimeric mucosal IgA is synthesized by intestinal lymphocytes. This IgA is transported across the epithelium via a secretory piece to the intestinal lumen.

The organized gut-associated lymphoid tissue consists primarily of lym- phoid patches named after the Swiss scientist Johann Conrad Peyer [l653– l712] who in 1677 described these patches as "intestinal glands." Peyer's patches primarily contain CD4+ T cells which are able to produce the cytokines IL4, IL5 and IL6 which allow B cells to switch to IgA immunoglobulin production.[3] Patients with ataxia telangiectasia and selective IgA deficiency, as well as AIDS patients, either lack, or have reduced numbers of, CD4+ cells in Peyer's patches.

Peyer's patches are covered with a dome-like membrane or M cells which take up intestinal organisms.[4] If the organisms lack virulence genes or viru- lence-inducing plasmids they are excluded, while those possessing such quali- ties are taken up and transported to the lamina propria and Peyer's patches. The antigenic determinants are processed by dendritic cells which in conjunction with histocompatibility molecules initiate CD4+ activity. These antigenically stimulated lymphocytes move to mesenteric nodes where they clone and differentiate. They then move into the thoracic duct, thence to the systemic circulation and home to the lamina propria. This latter portion of the mucosal immune system is often called the efferent limb and consists not only of the lymphocytes in the lamina propria but also of an intraepithelial component [IEL]. This less organized area consists of multiple cell types capable of inflammatory responses which include T and B lymphocytes, macrophages, neutrophils, eosinophils, mast cells and fibroblasts. Since most T cells in this efferent limb are activated by previous antigenic exposure, this compartment is crammed with cytokine-producing cells. The cells in the IEL consisting mainly of CD8+ T cells and both T-$\alpha\beta$ and T-$\gamma\delta$ receptors, the former capable of suppression and cytotoxicity and the latter of contra-suppression.[3-7] This latter activity may be responsible for elevated serum IgA immunoglobulin levels found in most patients with HIV infection.

In patients infected with HIV and gastrointestinal complaints, CD8+ T cells are increased in the lamina propria and IEL, while the CD4+ T cells in Peyer's patches are reduced.[8,9] Consequently, not only *Salmonella* spp, but also other virulent pathogens such as *Shigella* spp, *Campylobacter jejuni* and Campylobacter-like organisms, *Yersinia enterocolitica* and *Listeria monocytogenes,* can gain access to the regional mesenteric nodes and systemic circulation leading to bacteremia.[10-12] Diarrhea and dysentery result from damage to M cells and intestinal enterocytes by these virulent organisms.

Since these enteric bacterial infections occur early in the course of HIV infection, CD4+ T cell counts are usually not severely depressed and appropriate antimicrobial therapy is effective if continued for at least 2 weeks.[13-15] Malabsorption and malnutrition may occur with these infections, but since the organisms are ultimately eradicated, long-term nutritional deficits are rarely seen.

We have recently had the opportunity to evaluate 5 patients with *Pseudomonas aeruginosa* bacteremia and HIV infection who were receiving trimethoprim/sulfamethoxazone chemoprophylaxis for *Pneumocystis carinii* pneumonia.[16] In all likelihood *P. aeruginosa* colonized the intestinal tract under antimicrobial pressure and, because of the immunologic mucosal defect, bacteremia due to uptake by M cells developed.

Salmonella arizonae, a pathogen of reptiles and amphibians, is an etiologic agent of bacteremia in Hispanic-Americans who utilize folk medicine including capsules containing the pulverized viscera of the rattlesnake to cure a multitude of ailments including AIDS.[17-19]

INTESTINAL INFECTIONS CAUSED BY MYCOBACTERIA

Disseminated infection by *Mycobacterium avium* was rare before the AIDS pandemic. Prior to 1982 less than 50 cases had been reported.[20-22] Since then it has become apparent that *M. avium* is the most common bacterial infection in AIDS patients in the USA, and disseminated disease is diagnosed antemortem in 18% to 34%[23-25] and at autopsy in 33% to 53%.[26-28] *M. avium* infections in Great Britain and in Denmark are less common. St Mary's Hospital in London reported a 17% incidence from all sights.[29] In contrast, *M. avium* infections are rarely seen in AIDS patients from Africa[30] where *M. tuberculosis* is much more common. In the USA, tuberculosis in AIDS patients is found principally in groups with recognized risk factors for human mycobacteriosis.

At our institution nearly 300 patients with AIDS and mycobacterial infection have been seen since 1981. Only six patients did not have *M. avium* as the isolate: four were due to *M. tuberculosis* and one each to *M. bovis* and *M. chelonae*. Three of the four with *M. tuberculosis* presented with lymphadenopathy and one with bilateral pulmonary infiltrates, while the patient with *M. bovis* developed meningitis and multiple abdominal abscesses. The *M. chelonae* infection was secondary to an infected venous access sight. The remaining 96% of our patients had infection caused by *M. avium;* no infection was traceable to *M. intracellulare.*

M. avium was initially classified as serotypes 1–3, while the remaining serotypes are designated *M. intracellulare*. Their close phenotypic similarity led Meissner et al.[31] to group them together as *M. avium-intracellulare* complex [MAC]. When MAC strains from AIDS patients were studied using DNA probes, almost all were *M. avium.*[32] Most were also found to be serotypes 4, 6 and 8 which now all classify as *M. avium*. It is of more than passing interest to note that *M. avium* is closely related to slow-growing, mycobactin dependent, *Mycobacteria paratuberculosis* and the "wood pigeon bacillus". The former is associated with Johne's disease, an intestinal infection of ruminants, and both have been linked to the etiology of regional ileitis or Crohn's disease of man.[33-35]

The three serotypes found almost exclusively in AIDS patients stand in contrast to serotypes 1,2 and 3 which characteristically cause disease in birds and animals. There is little doubt, however, that the source of the organism is the environment, and that soil, dust and water are commonly found to harbor *M. avium*. Recently it has been shown that serovars 1, 4 and 8 have enhanced virulence due to the presence of small and large plasmids.[36-38] However, only 5–6% of soil and dust specimens from endemic areas possessed plasmids, while 75% of aerosolized specimens from the same areas contained plasmids. Our hospital is in a large arid area of California known as the Colorado desert with a mean annual rainfall of less than 4 inches and very little surface water; yet as noted above it provides medical care for a large number of *M. avium*-infected AIDS patients.

M. avium infections in HIV-infected individuals usually occur when CD4+ T cell counts are below 100/mm^3. The disease presents with fever, weight loss, abdominal pain, diarrhea, anemia and malabsorption. Serum carotene levels are invariably depressed, D-xylose absorption is frequently impaired, and stool fat excretion is always elevated. *M. avium* is ordinarily isolated from blood but can also be found in bone marrow and liver biopsy cultures. Probably because of impaired cellular immunity, well-defined granulomas are generally not seen by microscopy. Macrophages packed tightly with acid-fast bacilli can be identified using special stains.

Hepatosplenomegaly and lymphadenopathy involving the mesenteric and retroperitoneal nodes are usually seen with computerized tomography of the abdomen. Upper endoscopy reveals extensive involvement of the duodenum, although the entire gastrointestinal tract can be involved. Because of the more severe and extensive gastrointestinal involvement compared to the lung, the former is felt to be the portal of entry in AIDS patients. Life-style is not particularly a predisposing factor since children and heterosexual males and females are at equal risk. Although most attention has been focused on *M. avium, M. tuberculosis* and *M. bovis* can also cause extensive gastrointestinal disease in HIV-infected patients. Hirschel et al.[39] recently reported an infection with a novel unidentified organism resembling *M. simiae* which failed to grow on solid media but showed limited multiplication in BACTEC® liquid medium. Böttger[40] tentatively named the new species *"M. genavense"* and described 18 cases in AIDS patients. Others[41-43] noted that the organism causes a disease that resembles *M. avium* infection.

A number of drugs are available with efficacy against *M. avium*. These include rifampin, ethambutol, clofazimine, ciprofloxacin, amikacin, streptomycin, rifabutin, clarithromycin and azithromycin. A combination of four drugs is usually used at our institution, and generally within 12 weeks most patients are clinically improved and have sterile blood cultures. However, the weight loss and anemia are not significantly better with treatment. Peak blood levels have been studied following oral administration of isoniazid, rifampin ethambutol and clofazimine,and more than 80% of these determinations were below the desired therapeutic range. Thus, malabsorption of these antimicrobials could contribute to the inadequate response to treatment.[44-46] In an effort to obviate this problem, we have given isoniazid, rifampin, amikacin and ciprofloxacin parenterally to those patients who fail to respond to oral therapy, and have met with better therapeutic success.

INTESTINAL INFECTIONS CAUSED BY PROTOZOA

Cryptosporidium parvum is an acid-fast coccidian parasite which is recognized as an important cause of diarrheal disease in livestock as well as in both immunologically competent and immunocompromised humans. The parasite is not invasive and attaches to the plasma membrane of epithelial cells. Light microscopy shows evidence of small intestinal injury including partial villous atrophy. While usually mild and self-limited in immunocompetent individuals, patients with HIV infection often suffer protracted and life-threatening infections.

In the Western USA, high altitude streams very often are infested — probably due to contamination by feral animals. The parasite has a complex life cycle. After the thick-walled oocyst is ingested, sporozoites emerge and infect the gut epithelial cells. The trophozoite then forms a Type I meront which gives rise to merozoites which reenter the epithelium perpetuating the infection. Type II meronts are then formed which ultimately give rise to micro- and macrogametes that form a zygote, the oocyst. The thick-walled oocysts are excreted while the thin-walled oocysts remain in the host and release more sporozoites.[47]

Mechanisms of resistance and recovery from cryptosporidial infections are still not understood, although studies in mice suggest that T cell mechanisms are important for recovery.[48-50] Antibody alone has been found to prevent new infection in animals and man, and preincubation of sporozoites with monoclonal antibodies to sporozoite antigens reduced sporozoite infectivity 80–100% in a mouse model.[51-54] A >900,000-MW *Cryptosporidium parvum* sporozoite glycoprotein has been identified by Petersen et al.[55] which is recognized by protective hyperimmune bovine colostral immunoglobulin. Thus it seems likely that both CD4+ T cells and IgA mucosal antibody, probably directed at several antigens as well as non-specific mechanisms, are necessary mechanisms of resistance and recovery. It is of interest to note that similar mechanisms eliminate *Giardia lamblia*,[50] which, however, is not a significant pathogen in patients with HIV infection but is in individuals with hypogammaglobulinemia. This difference may, in part, be due to the complexity of the parasites respective life cycles.

Another acid-fast coccidian parasite causing diarrhea in HIV-infected patients is *Isospora belli*. Similar to cryptosporidiosis, the infection is most aggressive in immunocompromised patients. We have also seen sequential and then simultaneous infection by both parasites in a patient with AIDS. Since athymic rats do not develop resistance to this parasite it can be inferred that cell-mediated immunity plays a role in recovery from infection.

Finally, it is becoming evident that a complex group of parasites of the phylum *Microsporidia* which consists of about eighty genera is being detected with increasing frequency in patients with AIDS and chronic diarrhea. First identified in 1857, they cause economic problems for the silkworm, honeybee and commercial fishing industries. These are small and unicellular parasites, and cell division takes place at the sporont and meront stages. The most common species is *Enterocytozoon bieneusi* and others include *Encephalitozoon cuniculi, Nosema connori* and *Pleistophora*. With a variety of techniques including electron microscopy and Giemsa-stained smears, microsporidia have been identified in 14% to 30% of AIDS patients with chronic diarrhea.[56-58] More recently, studies from the Centers for Disease Control have shown good sensitivity and specificity with a special staining procedure using stool.[59]

All these parasites are capable of causing non-febrile chronic diarrhea and malabsorption with severe malnutrition accompanied by weight loss and abdominal discomfort. Serum carotene levels are decreased and D-xylose absorption is abnormal. Daily stool for fat shows significantly increased amounts and a Schilling test often may be impaired.

Although cryptosporidia, microsporidia and *Isospora belli* primarily invade the small bowel, all are capable of extra-luminal invasion to the hepatobiliary system, peritoneum, lung and cornea.

Therapy for *Isospora belli* has been quite satisfactory with trimethoprim/ sulfamethoxazole and pyrimethamine, although therapy must be continued for at least one to two months to prevent relapse. Cryptosporidiosis is rather recalcitrant to treatment, although paromomycin has been transiently effective in some patients with small bowel disease only. Since the drug is not absorbed it has no value for extra-intestinal disease. In our experience with diclazuril in 12 patients with HIV infection and cryptosporidiosis, a favorable response has been noted only occasionally. Spiromycin is generally felt to be ineffective therapy.

Patients with AIDS and microsporidiosis have been considered untreatable. However, recent reports[60,61] have shown that metronidazole is effective empiric therapy for symptomatic intestinal microsporidiosis, although such therapy does not eradicate the parasite. Dieterich et al.[62] have recently published favorable results with albendazole therapy.

The diagnoses of these parasitic infestations can all be made by stool examination, except for microsporidiosis which still requires small bowel biopsy unless the new stool examination technique proves to be as sensitive and specific.

VIRAL INFECTIONS OF THE GASTROINTESTINAL TRACT

Cytomegalovirus infections occur commonly in HIV+ individuals and disseminated disease is found in almost 90% of AIDS patients at autopsy. The spectrum of infection varies from asymptomatic excretion to invasive disease of the entire gastrointestinal tract, liver, lungs and eye. Gastrointestinal infection is associated with anorexia, nausea, vomiting and diarrhea. Additionally, acalculous cholecystitis, cholangitis, pancreatitis and adrenalitis are noted in patients with AIDS.

Colonic disease is the most common non-retinal expression of CMV infection. In nearly 30% of such infections, the main sight of pathologic change is in the cecum resulting in typhlitis. The clinical presentation resembles that seen in patients with neutropenic enterocolitis,[63,64] a syndrome found mainly in patients with hematologic malignancies who undergo chemotherapy, patients following transplantation, individuals who take non-steroidal anti-inflammatory drugs, or patients with uremia.[66] It appears likely that ileocecal ulceration is the fundamental lesion in neutropenic enterocolitis which may be due to the antineoplastic agents or other drugs or diseases; but in a number of cases this lesion has been due to CMV infection confined to the cecum.[67] Secondary sepsis occurs commonly in neutropenic enterocolitis and is often due to *Clostridia* spp and enterobacteria or fungi. However bacteremia does not occur with CMV typhlitis where right-sided abdominal pain, tenderness, fever and diarrhea are characteristic signs and symptoms. CT scanning of the abdomen shows thickening of the cecal wall, luminal narrowing and inflammatory changes in the surrounding mesentery and pericecal fat.[68-70] Barium studies often show edema, mucosal fold thickening and superficial erosions to deep ulcerations or even mass lesions. Gallium-56 nuclear scanning or, preferably, Indium-111 have also been used to demonstrate increased cecal uptake.

Most often, however, CMV disease involves other sites in the colon. Although diagnosis can occasionally be established by isolating the virus from blood, most often tissue needs to be harvested and the virus identified either by characteristic inclusions or by DNA hybridization probes. Colonoscopy provides the technique by which appropriate tissue can easily be obtained. Since CMV infection in AIDS patients is usually associated with reactivation of a latent virus rather than a primary infection, serologic studies are not helpful. IgG antibody titers are generally elevated while IgM levels remain insignificant or undetectable.

Treatment of CMV infections has been refashioned since the introduction of ganciclovir and foscarnet (tri-sodium phosphofonate). Both agents are effective in the management of CMV retinitis and gastrointestinal disease in patients with AIDS.[71] Unfortunately, both drugs are administered intravenously with neutropenia and renal toxicity respectively as more than occasional side effects. Long-term therapy is necessary, and probably some type of maintenance treatment is required to prevent relapse.

In general, fever, abdominal pain, diarrhea and nutritional status all improve following successful antiviral therapy; however, in some studies the results are somewhat inconsistent.[72] Although malabsorption in CMV colitis is not usual, simultaneous ileitis due to CMV occasionally occurs resulting in fat malabsorption and abnormal D-xylose studies. Additionally, weight loss could be due to the systemic effects of disseminated disease, and successful therapy could lead to a beneficial change in nutritional status.

Janoff et al.[73] recently reported evidence of adenoviral diarrhea in patients with AIDS. The virus was demonstrated by electron microscopy and/or by culture. These findings were confirmed by Hierholzer et al.[74] from Australia who also found evidence of rotavirus infection. A similar study,[75] however, failed to confirm these associations.

MISCELLANEOUS PARASITIC INFECTIONS

The term gay bowel syndrome was introduced in the 1960's to describe intestinal disease in homosexual males from whom *Entamoebae* and *Giardia lamblia* and a variety of enteric bacteria were isolated. With the exception of *Giardia,* these intestinal parasites as well as *Blastocystis hominis* appear to be uncommonly associated with disease in patients with AIDS. The clinical course of giardiasis is, however, no different in AIDS patients than in those not infected with HIV.

Finally, a recent report of diarrhea in an AIDS patient due to a blue-green alga[76] which can be identified as a large Cryptosporidium by acid-fast staining of stool adds yet another group of opportunistic pathogens to those causing intestinal disease in HIV+ persons.

There remains, however, a population of patients (15–30%) who are HIV+ and have diarrhea for more than 30 days and in whom a thorough investigation fails to yield a microbiologic etiology. Such patients very often display an abnormal small bowel biopsy demonstrating villous atrophy and increased CD8+ lymphocytes in the lamina propria. In addition, such individuals often show signs of weight loss and malabsorption as evidenced by a low serum carotene, abnormal D-xylose absorption and an abnormal Schilling test indicative of vitamin B_{12} malabsorption. These patients tend to have less than a 5 kg weight loss and stool volumes are generally less than 400 ml per day. At the present time such HIV+ persons are thought to have either HIV+ related diarrhea or their symptoms are due to bacterial overgrowth in the small bowel. *In situ* hybridization studies, immunohistochemical techniques or viral culture[77,78] have demonstrated evidence of HIV infection of the intestinal tissue. Whether the presence of HIV in the gut tissue truly represents disease-producing infection or asymptomatic infection remains to be determined. If indeed HIV causes symptomatic intestinal disease later in the course of AIDS, then the mechanisms will need to be classified.

The loss of the "gastric barrier" due to gastric secretory failure and subsequent small bowel bacterial overgrowth has been proposed as another cause of pathogen-unrelated diarrhea.[79-80] High fasting gastric pH has been described in a significant number of patients with AIDS. In some instances this secretory failure could be caused by *Helicobacter pylori*.

Although intestinal disease is most often associated with diarrhea in patients with AIDS, recent studies have shown a significant incidence of pancreatic disease at autopsy which can lead to symptoms of frequent, bulky and oily stools, glucose intolerance and involuntary weight loss. Since HIV has not been shown to infect acinar, islet or ductal cells of the pancreas, drug toxicities, opportunistic infections and neoplasia are the likely mechanisms for destruction of these elements.

Pentamidine, didanosine, and stavudine are drugs used in HIV+ patients that may cause pancreatitis along with opportunistic infections such as cytomegalovirus, toxoplasma, *Cryptococcus Candida* and *Mycobacteria*. Malabsorption due to intestinal disease is differentiated from maladigestion due to pancreatic disease by a D-xylose test or by a bentiromide test. Lack of pancreatic enzymes has no effect on D-xylose absorption, but in the absence of chymotrypsin, little para-aminobenzoic acid is produced from bentiromide to be absorbed and released in the urine.

The treatment of diarrhea and malnutrition of HIV+ persons is complex and can be costly. Simple measures such as diphenoxylate and similar commercially available compounds are sometimes helpful. At times tincture of opium or bismuth compounds are successful therapeutically. We have observed some efficacy with the somatostatin analogue octreotide acetate in patients with intractable diarrhea. Frequently none of the above measures is helpful.

Inadequate caloric intake can be offset by such appetite stimulants as megesterol acetate. Doses of 400 to 800 mg daily very often lead to significant weight gain in our experience. However, much of the gain in weight is due to fat accumulation. Newer studies with human growth hormones may result in increased muscle mass as well. The use of parenteral hyperalimentation is often helpful as an interim measure in correcting malnutrition.

We have learned much in a decade of AIDS. With time, many of the causes of malnutrition due to diarrhea will be understood and treatable, but cure of these illnesses will likely depend upon an improvement in the defective immunologic function.

REFERENCES

1. Serwadda D, Mugerwa RD, and Sewankambo NK. Slim disease: A new disease in Uganda and its association with HTLV-III infection. *Lancet* 1985;849–52.
2. Cone LA, Woodard DR, Potts BE, Byrd RG, Alexander RA, and Last MD. An update on the acquired immunodeficiency syndrome (AIDS). Associated disorders of the alimentary tract. *Dis Colon Rectum* 1986;29:60–4.
3. Targon SR. The lamina propria; A dynamic complex mucosal compartment. *Ann NY Acad Sci* 1992;664:61–8.
4. Owen RL and Jones AZ. Epithelial cell specialization within human Peyer's patches: An ultrastructural study of intestinal lymphoid follicles. *Gastroenterology* 1974;66:189–203.
5. Rosner AJ and Keren DF. Demonstration of M-cells in the specialized follicle-associated epithelium overlying isolated lymphoid follicles in the gut. *J Leukocyte Biol* 1984;35:397–404.
6. Brandtzaeg P, Halstensen TS, Kett K, Krajci P, Kvale D, Rognum TC, Scott H, and Sollid LM. Immunobiology and immunopathology of human gut mucosa: humoral immunity and intraepithelial lymphocytes. *Gastroenterology* 1989;97:1562–84.
7. McGhee JR, Mestecky J, Elson CO, and Kiyono H. Regulation of IgA synthesis and immune response and interleukins. *J Clin Immunol* 1989;9:175–99.
8. Rodgers VD, Fassett R, and Kagnoff MF. Abnormalities in intestinal mucosal T-cells in homosexual populations including those with lymphadenopathy syndrome and acquired immunodeficiency syndrome. *Gastroenterology* 1986;90:552–8.
9. Budhraja M, Levendoglu K, Kocka F, Mangkornkanos KN, and Sherer R. Duodenal mucosal T-cell subpopulation and bacterial cultures in acquired immune deficiency syndrome. *Am J Gastroenterol* 1987; 82:427–31.

10. Wassef JS, Keren DF, and Mailloux JL. Role of M-cells in initial antigen uptake and in ulcer formation in the rabbit intestinal loop model of Shigellosis. *Infect Immunol.* 1989;57:858–63.

11. Walker RI, Schmauder-Chock EA, and Parker JL. Selective association and transport of *Campylobacter jejuni* through M-cells of rabbit Peyer's patches. *Can J Microbiol* 1988;34:1142–7.

12. Grützkau A, Hanski C, Hahn H, and Riecken EO. Involvement of M-cells in the bacterial invasion of Peyer's patches: A common mechanism showed by *Yersinia enterocolitica* and other enteroinvasive bacteria. *Gut* 1990;31:1011–15.

13. Dworkin B, Wormser GP, Abdoo RA, Cabello F, Aguero ME, and Sivak SL. Persistence of multiple antibiotic resistant *Campylobacter jejuni* in a patient with acquired immunodeficiency syndrome. *Am J Med* 1986;80:965–70.

14. Perlman DM, Ampel NM, Schifman RB, Cohn DL, Patton CM, Aguirre ML, Wang WLL, and Blaser MJ. Persistent *Campylobacter jejuni* infections in patients infected with human immunodeficiency virus (HIV). *Ann Intern Med* 1988;108:540–6.

15. Bernard E, Roger PM, Carles D, Bonaldi V, Fournier JP, and Dellamonica P. Diarrhea and *Campylobacter* infections in patients infected with the human immunodeficiency virus (letter). *J Infect Dis* 1989;159:143–4.

16. Byrd RG, Cone LA, Woodard DR, Schulz P, and Stone RA. Community-acquired *Pseudomonas aeruginosa* bacteremia in patients with AIDS (Abst). *Clin Res* 1993, 108A.

17. Cone LA, Boughton WH, Cone LA, and Lehv LH. Further observations on rattlesnake capsule induced *Salmonella arizonae* bacteremia. *West J Med* 1990;153:315–6.

18. Casner PR and Zuckerman MJ. *Salmonella arizonae* in patients with AIDS along the Mexican border (letter). *N Engl J Med* 1990;323:198–199.

19. Waterman SH, Juarez G, and Carr SJ. *Salmonella arizona (sic)* infections in latinos associated with rattlesnake folk medicine. *Am J Public Health* 1990;80:286–9.

20. Greene JB, Sidhu GS, Lewin S, Levine JF, Masur H, Simberkoff MS, Nicholas P, Good RC, Zolla-Pazner SB, Pollack AA, Tapper ML, and Holzman RS. *Mycobacterium avium-intracellularae:* A cause of disseminated life-threatening infection in homosexuals and drug abusers. *Ann Intern Med* 1982;97:539–46.

21. Zakowski P, Fligiel S, Berlin GW, and Johnson BL. Disseminated *Mycobacterium avium intracellularae* in homosexual men dying of acquired immunodeficiency. *JAMA* 1983;248:2980–2.

22. Fainstein V, Bolivar R, Mavligit G, Rios A, and Luna M. Disseminated infection due to *Mycobacterium avium-intracellularae* in a homosexual man with Kaposi's sarcoma. *J Infect Dis* 1982;145:586.

23. Macher AM, Kovacs JA, Gill V, Roberts GD, Ames J, Park CH, Straus S, Lane HC, Parillo JE, Fauci AS, and Masur H. Bacteremias due to *Mycobacterium avium-intracellularae* in the acquired immunodeficiency syndrome. *Ann Intern Med* 1983;99:782–85.

24. Keihn TE, Edwards FF, Brannon P, Tsang AY, Maio M, Gold JWM, Whimbey E, Wong B, McClatchy JK, and Armstrong D. Infections caused by *Mycobacteria avium* complex in immunocompromised patients: Diagnoses by blood culture and fecal examination, antimicrobial susceptibility tests and morphobiological and seroagglutination characteristics. *J Clin Microbiol* 1985;21:168–73.

25. Wallace JM and Hannah JB. *Mycobacterium avium* complex infection in patients with the acquired immunodeficiency syndrome. A clinicopathologic study. *Chest* 1988;93:926–32.

26. Reichert CM, O'Leary TJ, Levens DL, Simrel CR, and Macher AM. Autopsy pathology in the acquired immunodeficiency syndrome *Am J Pathol* 1983;112:357–82.

27. Klatt EC, Jensen DF, and Meyer PR. Pathology of *Mycobacterium avium-intracellulare* infection in acquired immunodeficiency syndrome. *Hum Pathol* 1987;18:709–14.

28. Sohn CC, Schroff RW, Kliewer KE, Lebel DM, and Fligiel S. Disseminated *Mycobacterium avium-intracellularae* in homosexual men with acquired cell-mediated immunodeficiency: A histologic and immunologic study of two cases. *Am J Clin Pathol* 1983;79:247–52.

29. Coker RJ, Hellyer TJ, Brown IN, and Weber JN. Clinical aspects of mycobacterial infections in HIV infection. *Res Microbiol* 1992;143:377–381.

30. Okello DO, Sewankambo N, Goodgame R, Aisu TO, Kwezi M, Morrissey A, and Ellner JJ. Absence of bacteremia with *Mycobacterium avium-intracellularae* in Ugandan patients with AIDS. *J Infect Dis* 1990;162:208–10.

31. Meissner G, Schroeder KH, Amadio GE, *et al.* A cooperative numerical analysis of non-scoto-chromogenic slowly growing mycobacteria. *J Gen Microbiol* 1974;83:207–35.

32. McFadden JJ, Kunze ZM, Portaels F, Labrousse V, and Rastogi N. Epidemiological and genetic markers, virulence factors and intracellular growth of *Mycobacterium avium* in AIDS. *Res Microbiol* 1992;143:423–30.

33. Chiodini RJ. Crohn's disease and the mycobacteroses: A review and comparison of two disease entities. *Clin Microbiol Rev* 1989;2:90–117.

34. McFadden J, Collins J, Beaman B, Arthur M, and Gitnick G. Mycobacteria in Crohn's disease: DNA probes identify the wood pigeon strain of *Mycobacterium avium* and *Mycobacterium paratuberculoses* from human tissue. *J Clin Microbiol* 1992;30:3070–3.

35. Gitnick G, Collins J, Beaman B, Brooks D, Arthur M, Imaeda T, and Palieschesky M. Preliminary report on isolation of Mycobacteria from patients with Crohn's disease. *Dig Dis Sci* 1989;34:925–32.

36. Crawford JT and Bates JH. Analysis of plasmids in *Mycobacterium avium-intracellularae* isolates from persons with acquired immunodeficiency syndrome. *Am Rev Respir Dis* 1986;134:659–61.

37. Gangadharam PRJ, Perumal VK, Crawford JT, and Bates JH. Association of plasmids and virulence of *Mycobacterium avium* complex. *Am Rev Resp Dis* 1988;137:212–14.

38. Meissner PS and Falkingham JO III. Plasmid DNA profiles as epidemiological markers for clinical and environmental isolates of *Mycobacterium avium*, *Mycobacterium intracellularae* and *Mycobacterium scrofulareum*. *J Infect Dis* 1986;153:325–31.

39. Hirschel B, Chang HR, Mach N, Piguet P, Cox J, Piguet JD, Silva MT, Larsson L, Klatser PR, Thote ER, Rigaits L, and Portaels F. Fatal infection with a novel unidentified mycobacterium in a man with the acquired immunodeficiency syndrome. *N Engl J Med* 1990;323:109–13.

40. Böttger EC, Teske A, Kirscher P, Bost S, Chang HR, Beer V, and Hirschel B. Disseminated *"Mycobacterium genavense"* infection in patients with AIDS. *Lancet* 1992;340:76–80.
41. Wald A, Coyle MB, Carlson LC, Thompson RL, and Hooton TM. Infection with a fastidious Mycobacterium resembling *Mycobacterium semiae* in seven patients with AIDS. *Ann Intern Med* 1992;117:586–9.
42. Jackson K, Sievers A, Ross BC, and Dwyer B. Isolation of a fastidious *Mycobacterium* species from two AIDS patients. *J Clin Microbiol* 1992;30:2934–7.
43. Coyle MB, Carlson LC, Wallis CK, Leonard RB, Raisys VA, Kilburn JO, Samadpour M, and Böttger EC. Laboratory aspects of *"Mycobacterium genevense"* a proposed species isolated from AIDS patients. *J Clin Microbiol* 1992;30:3206–12.
44. Cone LA, Woodard DR, Fiala M, Tornay AS, and Byrd RG. Mycobacterial disease in AIDS patients residing in the Colorado Desert of California (abst). VII International Conf on AIDS, Florence, Italy, June 16–21, 1991. TH.C.94.
45. Berning SE, Huitt GA, Iseman MD, and Peloquin OA. Malabsorption of antituberculous medications by a patient with AIDS (letter). *N Engl J Med* 1992;327:1817–8.
46. Peloquin OA. Therapeutic drug monitoring: Principles and application in mycobacterial infections. *Drug Ther* 1992;22:31–6.
47. Current WL, Reese NC, Ernst JV, Bailey WS, Heyman MB, and Weinstein WM. Human cryptosporidiosis in immunocompetent and immunodeficiency persons — studies of an outbreak and experimental transmission. *N Engl J Med* 1983;308:1252–7.
48. Soave R, Danner RL, Honig CL, Ma P, Hart CC, Nash T, and Roberts RB. Cryptosporodiosis in homosexual men. *Ann Intern Med* 1984;100:504–11.
49. Crawford FG and Vermund SH. Human Cryptosporidiosis. *Crit Rev Microbiol* 1988;16:113–59.
50. Heyworth MF. Immunology of *Giardia* and *Cryptosporidium* infections. *J Infect Dis* 1992;166:465–72.
51. Heine J, Moon HW, and Woodmansee DB. Persistent *Cryptosporidium* infection in congenitally athymic (nude) mice. *Infect Immun* 1984;43:856–9.
52. Ungar BLP, Burris JA, Quinn CA, and Finkelman FD. New mouse models for chronic *Cryptosporidium* infection in immunodeficient hosts. *Infect Immun* 1990;58:961–9.
53. Whitmire WM and Harp JA. Characterization of bovine cellular and serum antibody response during infection by *Cryptosporidium parvum*. *Infect Immunol* 1991;59:990–5.
54. Riggs MW, McGuire TC, Mason PH, and Perryman LE. Neutralization-sensitive epitopes are exposed on the surface of infectious *Cryptosporidium parvum* sporozoites. *J Immunol* 1989;143:1340–5.
55. Petersen C, Gut J, Doyle PS, Crabb JH, Nelson RG, and Leech JH. Characterization of a >900,000-MW *Cyptosporidium parvum* sporozoite glycoprotein recognized by protective hyperimmune bovine colostral immunoglobulin. *Infect Immun* 1992;60:5132–8.
56. Bartlett JG, Belistos PC, and Seers CL. AIDS enteropathy. *Clin Infect Dis* 1992;15:726–35.

57. Kotler DP, Francisco A, Clayton F, Scholes JV, and Orenstein JM. Small intestinal injury and parasitic diseases in AIDS. *Ann Intern Med* 1990;113:444–9.

58. Orenstein JM, Chiang J, Steinberg M, Smith PD, Rotterdam H, and Kotter DP. Intestinal microsporidiosis as a cause of diarrhea in human immunodeficiency virus-infected patients: a report of 20 cases. *Hum Pathol* 1990;21:475–81.

59. Weber R, Bryan RT, Owen RL, Wilox M, Gorelkin L, and Visvesvara GS. Improved light-microscopical detection of microsporidia spores in stool and duodenal aspirates. The Enteric Opportunistic Infections Working Group. *N Engl J Med* 1992;326:161–6.

60. Eeftinck Schattenkerk JK, vanGool T, vanKetel RJ, Bartelsman JFWM, Kuiken CL, Terpstra WJ, and Reiss P. Clinical significance of small-intestinal microsporidiosis in HIV-1 infected individuals. *Lancet* 1991;337:895–8.

61. Beaugerie L, Teilhac M-F, Deluol A-M, Fritsch J, Girard P-M, Rozenbaum W, Le Quintrec Y, and Chatelet F-P. Cholangiopathy associated with *Microsporidia* infection of the common bile duct mucosa in a patient with HIV infection. *Ann Intern Med* 1992;117:401–2.

62. Dieterich A, Word S, Nieves DP, and Kotler DP. Albendazole treatment of microsporidial enteritis in AIDS patients caused by two different species (abst 1011). *Proceedings Digestive Diseases Week* 1992 (May).

63. Steinberg D, Gold J, and Brodin A. Necrotizing enterocolitis in leukemia. *Arch Intern Med* 1973;131:538–44.

64. Mower WJ, Hawkins JA, and Nelson EW. Neutropenic enterocolitis in adults with acute leukemia. *Arch Surg* 1986;121:571–4.

65. Alt B, Glass NR, and Sollinger H. Neutropenic enterocolitis in adults. Review of the literature and assessment of surgical intervention. *Am J Surg* 1985;149:405–8.

66. Last MD and Lavery IC. Major hemorrhage and perforation due to a solitary cecal ulcer in a patient with end-stage renal failure. *Dis Colon Rectum* 1983;25:586–8.

67. Wolfe BM and Cherry JD. Hemorrhage from cecal ulcers of cytomegalovirus infections. *Ann Surg* 1973;177:490–4.

68. Cho S-R, Tisnado J, Lui C-I, Beachley MC, Shaw C-I, Kipreos BE, and Schneider V. Bleeding cytomegalovirus ulcers of the colon: Barium enema and angiography. *AJR* 1981;135:1213–5.

69. Balthazar EJ, Megibow AJ, Fazzini E, Opulencia JF, and Engel I, Cytomegalovirus in AIDS: Radiographic findings in 11 patients. *Radiology* 1985;155:585–9.

70. Duva-Frissora AD. Cytomegalovirus pseudotumor of the cecum. *AJR* 1991;156:1302–4.

71. Kotler DP. Cytomegalovirus colitis and wasting. *J Acquir Immune Defic Synd* 1991;4 (suppl 1);536–41.

72. Reed EC, Wolford JL, Kopecky KJ, Lilleby KE, Dandliker PS, Todaro JL, McDonald GB, and Meyers JD. Gancyclovir for the treatment of cytomegalovirus gastroenteritis in bone marrow transplant patients. *Ann Intern Med* 1990;112:505–10.

73. Janoff EN, Orenstein JM, Manischewitz JF, and Smith PD. Adenovirus colitis in the acquired immunodeficiency syndrome. *Gastroenterology* 1991;100:976–9.

74. Hierholzer J, Wigand R, Anderson LJ, Adrian T, and Gold JWM. Adenoviruses from patients with AIDS: a plethora of serotypes and a description of five new serotypes of subgenus D (types 43–47). *J Infect Dis* 1988;158:804–813.
75. Smith PD, Lane HC, Gill VJ, Manischewitz JF, Quinnan GV, Fauci AS, and Mazur H. Intestinal infections in patients with the acquired immunodeficiency syndrome (AIDS). Etiology and response to therapy. *Ann Intern Med* 1988;108:328–33.
76. Long EG, White EH, Carmichael WW, Quinlisk PM, Raja R, Swisher BL, Daugharty H, and Cohen MT. Morphology and staining characteristics of a cyanobacterium-like organism associated with diarrhea. *J Infect Dis* 1991;164:199–202.
77. Fox CH, Kotler D, Tierney A, Wilson CS, and Fauci AS. Detection of HIV-1 RNA in the lamina propria of patients with AIDS and gastrointestinal disease. *J Infect Dis* 1989;159:467–71.
78. Nelson JA, Wiley CA, Reynolds-Kohler C, Reese CF, Margaretten W, and Levy JA. Human immunodeficiency virus detected in bowel epithelium from patients with gastrointestinal symptoms. *Lancet* 1988;1:259–62.
79. Lake-Bakaar G, Quadros E, Beidas S, Elsakr M, Tom W, Wison DE, Dincsoy HP, Cohen P, and Straus EW. Gastric secretory failure in patients with acquired immunodeficiency syndrome (AIDS). *Ann Intern Med* 1988;109:502–4.
80. Belistos PC, Greenson JK, Sisler J, Yardley JH, and Bartlett JG. Association of chronic wasting diarrhea with gastric hypoacidity and opportunistic enteric infections in patients with acquired immunodeficiency syndrome (AIDS). *J Infect Dis* 1992;166:277–84.

Chapter

7

Vitamins in HIV Infection

Gregg O. Coodley

Vitamins appear to have considerable clinical importance in HIV infection. Certain vitamins are frequently deficient in HIV-infected patients. Vitamin deficiency appears to be related to altered immune, hematologic and neurologic status in HIV-infected patients. This chapter will review current information about vitamins and their potential roles in HIV infection.

VITAMIN B$_{12}$ (COBALAMIN)

Many studies have demonstrated that vitamin B$_{12}$ deficiency occurs commonly in HIV-infected patients and may occur at any stage of infection.[1-28] Herbert and colleagues suggested that serum B$_{12}$ levels may underestimate the degree of deficiency in AIDS. They reported that serum levels of the delivery protein holotranscobalamin II and the amount of B$_{12}$ bound to it were a more sensitive marker of early deficiency and suggested that functional B$_{12}$ deficiency may occur in almost half of HIV-infected patients.[8,23-25]

The majority of studies have suggested that B$_{12}$ deficiency results from malabsorption.[5,6,10,11,20] Harriman et al studied eleven AIDS patients with low (three) or normal (eight) B$_{12}$ levels.[5] Eight of eleven had abnormal Schilling tests even when given both intrinsic factor and pancreatic enzyme supplements. Intestinal biopsies in these patients revealed histopathologic evidence of chronic inflammation with evidence that HIV-1 virus was present in the lamina propria. The investigators concluded that malabsorption, perhaps secondary to a direct inflammatory effect of the HIV-1 virus, may be the principal cause of B$_{12}$ deficiency in HIV infection.[5] Zeitz reported that 72% of patients tested had abnormal Schilling tests.[20] Similarly, Remacha reported that all six patients studied with B$_{12}$ deficiency and HIV infection had abnormal Schilling tests, showing evidence of ileal malabsorption.[10,11] El Sakr reported that intrinsic factor secretion is reduced in patients with AIDS in association with

0-8493-7842-7/94/$0.00+$.50
© 1994 by CRC Press, Inc.

decreased gastric acid secretion, suggesting parietal cell dysfunction or destruction.[6] Herbert suggested that B_{12} deficiency at the cellular level may be exacerbated by the deficiency of the delivery protein holotranscobalamin II (TC II) that occurs early in B_{12} deficiency, although Hansen et al reported that low plasma B_{12} was not simply due to low concentrations of B_{12} binding proteins.[8,9]

Another potential mechanism of B_{12} deficiency may be an effect of zidovudine (AZT).[26,27] Richman et al reported in a double blind study of AZT in HIV infection that AZT recipients had a higher frequency of reduction in B_{12} levels than patients taking placebo. Twenty patients (14%) given AZT, compared to eight (6%) taking placebo had at least one B_{12} level less than 200 pg/ml.[27] Falutz reported similar findings.[29] The mechanism of how zidovudine could act to decrease B_{12} levels is unknown.

The extent that vitamin B_{12} deficiency contributes to the anemia found in HIV infection is unclear.[29] Vitamin B_{12} replacement has been shown to lead to decreased anemia in cases of documented B_{12} deficiency in HIV infection.[11]

Vitamin B_{12} deficiency in HIV infection has been linked variably to neutropenia.[27,29] Richman et al, in a study of zidovudine, noted that lower serum B_{12} levels, even within the normal range, were independently correlated with increased development of neutropenia in patients.[27] However, in another study, Falutz did not find any correlation between B_{12} level and WBC or granulocyte count.[29] Baum et al reported in a 3-year longitudinal study of 111 HIV-1 seropositive males (CDC stage III at baseline) that vitamin B_{12} levels predicted change in the CD4 count and that normalization of B_{12} levels appeared to increase the number of CD4 cells.[28]

Three large studies have looked at whether prophylactic B_{12} would lessen the hematologic toxicity of zidovudine.[30-32] All three studies failed to show any benefit from prophylactic B_{12} in preventing anemia or increasing tolerance to zidovudine.[30-32]

Vitamin B_{12} deficiency has also been associated with neurologic dysfunction in HIV infected patients.[12,15] Baum et al administered cognitive function tests to 100 asymptomatic HIV-seropositive males. Patients with clear or marginal B_{12} deficiency had poorer cognitive function than patients with normal B_{12} levels. Moreover, those patients whose B_{12} levels normalized with intramuscular (IM) B_{12} therapy also experienced improvement in subsequent cognitive testing. Patients whose B_{12} levels did not normalize with IM B_{12} therapy had no significant improvement in cognition. The authors concluded that B_{12} deficiency was associated with cognitive impairment and that correction of the deficiency could result in functional improvement.[12,15] Similarly, Kiebwurtz et al reported that HIV-infected patients referred for neurological problems frequently had vitamin B_{12} deficiency, and vitamin B_{12} treatment resulted in clinical improvement.[13] In contrast, Stern et al reported that in asymptomatic HIV disease, neurocognitive deficiency did not correlate with B_{12} level.[14]

Other authors have speculated whether the HIV-associated condition spinal vacuolar myelopathy was secondary to B_{12} deficiency, noting that the same region of the spinal cord affected in this condition were those affected by B_{12} deficiency.[13,33] This hypothesis has not been proven at this point.

In summary, vitamin B_{12} deficiency is common in all stages of HIV infection, although its prevalence varies considerably among studies. The major cause of B_{12} deficiency appears to be malabsorption. B_{12} deficiency may contribute to cognitive impairment in HIV infection and its correction may lead to improved cognition. B_{12} deficiency is also a treatable cause of anemia and possibly neutropenia in HIV-infected patients.

VITAMIN B_6

Pyridoxine (vitamin B_6) deficiency has also been reported to be common in patients with HIV infection, with prevalences ranging from 12% to 52% of patients in a variety of studies.[4,17,22,28,34-36]

The pathogenesis of pyridoxine deficiency in AIDS is unclear. Mantero-Atienza and colleagues reported that pyridoxine deficiency occurred despite apparently adequate dietary intake, although they also noted that patients who took at least 20 mg of B_6 daily were able to avoid deficiency.[34-36] In contrast, Coulston et al suggested that HIV patients may have inadequate dietary intake of vitamin B_6. They found that vitamin B_6 intake was below the recommended daily allowance (RDA) in a group of 26 patients with early HIV disease.[37]

Prior animal and human studies have suggested that B_6 deficiency results in impairment of both cell-mediated and humoral immune responses including impaired interleukin-2 production and lymphocyte proliferation in response to mitogens.[38,39] In one study, Mantero-Atienza and colleagues reported that HIV-infected patients with vitamin B_6 deficiency had uniformly lower CD4 cell counts than those without B_6 deficiency.[34] In a later investigation, the authors gave 20–25 mg pyridoxine a day to 12 HIV-positive patients who were asymptomatic except for lymphadenopathy. Of the 12, 8 had significant increases in their CD4 cell counts (average change 121 + 31 CD4 cells/mm^3, range 7–291 cells/mm^3) over a 6-month period, with parallel changes in response to phytohemagglutinin, a test of functional immune status.[36] Two other reports suggested that vitamin B_6 deficiency in HIV-infected patients was associated with reduced natural killer cell cytotoxicity and decreased lymphocyte mitogen responsiveness.[28,35]

One study of 108 HIV positive patients suggested that B_6 deficiency occurred commonly (in 30%) and was significantly correlated with tension/anxiety and bipolar manic behavior and that correction of B_6 deficiency resulted in decreased depression.[40]

In conclusion, vitamin B_6 deficiency has been identified commonly in HIV infection, and may result in immunologic and psychiatric impairment. One study suggests that B_6 supplementation may raise CD4 cell counts, but further studies are needed to confirm the benefits of vitamin B_6 therapy.[36]

THIAMINE

While Beach et al and Bogden et al failed to find any cases of thiamine deficiency in their studies of 50 and 30 patients, respectively, Malcolm et al reported that 2 of 16 patients with AIDS/ARC had low thiamine levels and Butterworth reported that 9 of 39 (23%) of patients with AIDS or ARC had biochemical evidence of thiamine deficiency using the erythrocyte transketolase activation assay.[16-18,41]

There are three case reports of Wernicke's encephalopathy in AIDS patients, and a fourth case of acute encephalitis treated with AZT, dexamethasone and thiamine with rapid clinical resolution.[42-45] These cases have led the authors to postulate that thiamine deficiency may be more common than has been reported. Butterworth et al argued that thiamine supplementation should be started in all new cases of AIDS or ARC.[18] Since thiamine deficiency in non-HIV-infected patients is usually secondary to malnutrition, it is likely that malnutrition is the etiology in HIV-infected patients as well.

RIBOFLAVIN AND NIACIN

There is limited data on whether riboflavin or niacin deficiency are clinically important in HIV-infected patients. While 2 studies failed to find any patients with riboflavin deficiency, a third reported that 27% of 100 asymptomatic HIV-positive males had low levels of riboflavin.[17,22,41] In the only reported study of niacin, Bogden et al found that 7% of 30 HIV-infected patients at various stages of disease had serum niacin levels below the normal range.[17] This finding was not clearly correlated with stage of disease and the significance of niacin and riboflavin deficiency in HIV infection is unclear.[17]

VITAMIN A

Vitamin A deficiency occurs commonly in HIV infection. Eight of nine studies of HIV-infected patients found evidence of vitamin A deficiency with frequencies ranging from 5% to 29% of patients.[4,16,17,22,41,46-48] A number of interesting associations has also been noted. Javier reported that vitamin A deficiency in a group of asymptomatic patients correlated with decreased natural killer (NK) cells and levels of IgG.[46] A study of HIV-infected mothers in Rwanda reported that low maternal serum retinol levels were correlated with increased fetal, neonatal and postnatal death.[47] Coodley et al reported that serum vitamin A levels were significantly lower in patients with wasting.[4] Malabsorption secondary to HIV-induced intestinal dysfunction appears to be the likely etiology of vitamin A deficiency, but no studies about pathogenesis have been reported.

Prior to the HIV epidemic, numerous studies, from cell cultures, animals and humans, related vitamin A to the risks of infection. These studies showed increased rates and severity of infection when vitamin A is deficient.[49-52] They also demonstrated that vitamin A repletion or prophylactic administration may reduce the incidence or severity of infection.[49,50] Therefore, vitamin A deficiency could further predispose HIV-infected patients to more frequent or severe infections.

There have been two reports of vitamin A therapy in HIV infection. Schofer et al treated oral hairy leukoplakia in an HIV-infected patient with local application of .1% vitamin A twice daily. The complete resolution of the lesions after 10 days of therapy suggested that topical vitamin A may have antiviral properties.[53] Watson et al infected mice with the LP-BMS murine leukemia virus, producing an "AIDS-like condition".[54] They treated half of the mice with high dietary vitamin A (retinal palmitate) and half without this supplement. Of the vitamin A supplemented mice, 71% survived compared to 45% of controls.[54] The treated animals also had increased numbers of macrophages and total T cells.[54]

In addition, a number of studies now suggest that beta carotene, a carotenoid with provitamin A (retinol) activity, may have clinical efficacy in HIV infection. Uncontrolled trials of beta carotene, ranging from 60–180 mg a day, variably resulted in increased numbers of CD4 cells and natural killer cells.[55-57] Coodley et al conducted a placebo-controlled trial of beta carotene 180 mg a day vs placebo and found that the beta carotene resulted in an increased percent change in CD4/CD8 ratio, total WBC count and a trend to increased number of CD4 cells compared to placebo.[58] The mechanism of beta carotene's beneficial effect remain unclear.[58]

In summary, the small available literature demonstrates that vitamin A deficiency occurs in HIV infection and likely increases with more advanced disease. Limited data also suggest the possibility that vitamin A and/or beta carotene may be useful therapeutic agents in HIV-infected patients.

VITAMIN C

There is limited information about vitamin C in HIV disease. The four studies that report on vitamin C deficiency in HIV infections report prevalences that include 0%, 7%, 14% and 27% of patients.[4,17,41,46] Studies that included patients with more advanced disease tended to find more deficiency, although serum levels did not always correlate with stage of disease.[4,17] In one study, Javier reported that vitamin C deficiency correlated with decreased IgM levels.[46]

The benefit of vitamin C therapy in HIV infection is not clear. Cathcart reported that patients treated with high doses of ascorbate (50–200 grams/per 24 hours) had clinical improvement and decreased symptoms, despite a lack of change in their CD4/CD8 ratio.[59] In three studies, Harekeh et al reported that ascorbic acid reduced HIV reverse transcriptase activity in HIV-infected T lymphocyte cell lines in vitro as well as inhibiting syncytia formation and reverse transcriptase activity of extracellular HIV virus.[60-62]

VITAMIN D

The two studies that looked for vitamin D deficiency in HIV-infected patients reported quite different results. Malcolm et al found normal vitamin D levels in all 14 patients with AIDS/ARC studied.[16] In contrast, Coodley et al reported that 17% of 47 patients at different stages of HIV infection had 25 OH vitamin D deficiency and that 10% had $1,25(OH)_2$ vitamin D deficiency.[4] In this study, lower vitamin D levels appeared to correlate with decreased CD4 count and possibly with wasting.[4]

Several in vitro studies have suggested that vitamin D may enhance HIV replication.[63-66] For example, Locardi noted that vitamin D resulted in increased macrophage differentiation and a marked increase in HIV infection.[63] Skolnick and colleagues reported in two studies that $1,25(OH)2D_3$ greatly enhanced HIV-1 replication in monocyte cell cultures and peripheral blood monocytes, suggesting that vitamin D might be one of the most potent enhancers of HIV replication.[65,66] In contrast, Rigby and colleagues reported that vitamin D added to monocyte cell culture reduced productive infection of cells by HIV-1 by 95%.[67]

Studies are also contradictory on the effect of vitamin D on immune responsiveness. While Girasole reported that vitamin D stimulated monocyte chemotaxis in serum of AIDS patients, Tobler et al reported that vitamin D may suppress GM-CSF expression in lymphocytes, thus further attenuating the normal immune response to infections.[68,69] These in vitro studies suggest that vitamin D may have an important, although as yet unclear, influence on the immune response, particularly that of T lymphocytes.

VITAMIN E

Four studies have been reported on the prevalence of vitamin E deficiency in HIV infection. Beach et al and Coodley et al reported normal vitamin E levels in all patients studied, while Bogden et al found that 12% of 30 patients, at varying stages of HIV infection, had low plasma levels of vitamin E.[4,17,41] Javier reported that 27% of 70 asymptomatic patients were deficient in vitamin E, although this did not correlate with changes in immune parameters.[46] The cause of vitamin E deficiency is unclear, although it may relate to the malabsorption associated with worsening HIV infection.

Several investigators have hypothesized that vitamin E may favorably modulate the immune response in HIV infection.[70-72] Odeleye and Watson noted that vitamin E has been shown to increase CD4/CD8 ratio, lymphocyte count, natural killer cell activity, phagocytosis and mitogen responsiveness[72] The authors reported that supraphysiological levels of vitamin E have been shown to increase immune responsiveness and increase host resistance to microorganisms.[72] Hollins argued that vitamin E stimulates the helper function of T cells, the mitogenesis of T cells and perhaps T and B cell cooperation.[70] Kline et al reported that vitamin E, studied in two animal retrovirus models, reduced retrovirus-induced T suppressor activity, increased interleukin-2 production and decreased PGE_2 production.[71]

Gogu and colleagues studied the effects of vitamin E on the HIV-1 virus in cell culture by the addition of vitamin E (in the form of alpha-D-tocopherol acid succinate [ATS]) with and without AZT, to HIV-1 infected human lymphocytic cell lines.[73] They found that the addition of vitamin E (ATS) alone had no significant effect. However, in combination with AZT it acted synergistically to inhibit the HIV-1 virus. The authors explained their results by postulating that ATS may modulate glycosylation of viral proteins and acts with AZT in creating a sequential blockade in HIV replication. The authors also noted that vitamin E (ATS) reduced the toxicity of AZT on bone marrow cells in culture and suggested that this might result either from a stimulating effect of vitamin E (ATS) on cellular proliferation or by activating erythropoietin.[73]

While these studies suggest that vitamin E may be beneficial in the treatment of HIV infection, actual trials of vitamin E in retrovirus-infected animals and HIV-infected patients are needed to confirm its potential utility.

VITAMIN K

The only available study of vitamin K and HIV infection reported that menaquinone (a bacterial vitamin K) suppressed HIV-1 induced syncytia formation in cell culture, but had no effect on virus production.[74] The authors postulated that if vitamin K could interfere with HIV induced syncytia formation, it might be useful in containing HIV-1 in vivo.[74]

FOLATE

There is marked disagreement about whether folate deficiency is a common clinical problem in HIV infection. Four studies reported that the majority of patients had normal or elevated folate levels and that folate deficiency was rare.[3,4,17,41]

In contrast, a number of studies have reported that folate deficiency occurs commonly in HIV-infected patients.[19-21,75] Boudes and colleagues evaluated folate levels in 74 HIV-infected patients and found that 64% of their patients not receiving folate replacement were folate deficient. Further, they found that only those patients receiving folate replacement had elevated serum folate levels. The authors concluded that folate deficiency may be common in HIV infection and that elevated serum levels likely reflect vitamin supplementation.[19] Herbert et al and Zeitz similarly reported that folate deficiency occurred in 66% and 41% of patients studied, respectively.[20,75] Revell reported that HIV-infected patients had significant impairment of absorption of folic acid, regardless of disease stage or degree of gastrointestinal symptoms, suggesting a possible mechanism of folate deficiency.[76]

Similarly, other researchers have suggested that the elevated serum folate levels observed in HIV infection may be spurious and be secondary to cellular destruction or increased immunoglobin binding in HIV infection leading to transient elevations of serum levels of folate.[77]

Two reports have suggested that cerebrospinal fluid (CSF) folate deficiency may occur in HIV infection and be a cause of neurologic dysfunction.[78-80]

The literature is also contradictory as to the benefit of folinic acid in preventing the hematologic toxicity of anti-folate drugs such as trimetrexate or trimethoprim in HIV infections.[81-85] Two studies of trimetrexate to treat PCP showed that leucovorin (folinic acid) minimized marrow toxicity and allowed the trimetrexate to be well tolerated.[81-83] Other investigators have argued that prophylactic folinic acid does not always prevent or reverse the cytopenia commonly observed in AIDS.[84,85] While there is no consensus about the role of folate in HIV infection, it appears to be a deficiency that can be measured easily and potentially corrected if present.

SUMMARY

It is important for clinicians to realize that HIV-infected patients may believe in the utility of vitamin supplementation and be taking one or more vitamin supplements, often without any consultation with their physician. Two studies have suggested that as many as half of patients may be taking vitamin supplements.[86,87] There is one uncontrolled report of the utility of multi-vitamin supplements. Priestley reported a series of 203 HIV-infected patients taking multi-vitamin supplements, in which this appeared to result in stabilization of CD4 counts and improved survival compared to survival of patients with similar CD4 counts in other studies.[88]

There is considerable evidence that vitamins are clinically important in HIV-infected patients. Deficiencies of folate, B_{12} and thiamine have been linked to neurological impairment. A number of vitamins, including vitamins A, B_6, C and E, may enhance (and their deficiency impair) the immune response in HIV-infected patients. While vitamin levels often correlate with nutritional intake, Baum et al reported that vitamin deficiencies of B_6, B_{12}, B_2, A, C and E may occur despite patients consuming vitamins at or above the recommended daily allowance (RDA).[89] Baum et al suggested that patients take levels of vitamins in excess of the RDA, particularly for these vitamins.[89]

Further research is needed to establish which deficiencies should be screened for in HIV-infected patients. In addition, further studies of the benefits of supplementation with particular vitamins, especially in terms of immune function, are needed.

Currently, clinicians may want to consider screening all HIV-infected patients for deficiencies of vitamins A, B_6, B_{12} and folate by measuring serum levels. This is despite the caveat that serum vitamin levels do not always reliably reflect nutrient status secondary to technical limitations. In patients with advanced AIDS, and particularly those with evidence of malnutrition and wasting, clinicians may consider screening more broadly for vitamin deficiency. While empiric vitamin supplementation conceivably could prevent development of deficiencies and their sequelae, such treatment has not yet been proven to prevent deficiency, and the issue of malabsorption raises questions about oral supplementation. Vitamins, particularly the fat soluble vitamins, also can cause toxicity, particularly when taken in large doses. Nevertheless, supplementation of all HIV-infected patients with a daily multivitamin appears to be a prudent step at this time.

Further recommendations must be based on the results of further research before they can be suggested for clinical practice.

ACKNOWLEDGMENTS

The author would like to acknowledge the tireless assistance of Ms. Darlene Coffey and Sue Simmons in the preparation of this manuscript.

REFERENCES

1. Burkes RL, Cohen H, Sinow RM, Levine AM, and Carmel R. Low serum B_{12} levels in homosexual males with AIDS or its prodrome. *Blood* 1984; 64 (Suppl 1):93a (abstract)
2. Burkes, RL, Cohen H, Krailo M, Sinow RM, and Carmel R. Low serum cobalamin levels occur frequently in the Acquired Immune Deficiency Syndrome and related disorders. *Eur J Haematol* 1987; 38:141–147.

3. Beach RS, Mantero-Atienza E, Eisdorfer C, et al. Altered folate metabolism in early HIV infection. *JAMA* 1988; 259:3129.
4. Coodley GO, Coodley MK, Nelson HD, and Loveless MO. Micronutrient concentrations in the HIV wasting syndrome. AIDS 1993, (in press).
5. Harriman G, Smith PD, Horne MK, et al. Vitamin B_{12} malabsorption in patients with Acquired Immunodeficiency Syndrome. *Arch Intern Med* 1989; 149:2039–2041.
6. El Sakr M, Ahuja F, and Lake-Bakaar G. Intrinsic factor and gastric acid secretion in HIV disease. Eighth Int Conf on AIDS. Amsterdam, July 19–24, 1992. Abstract PUB 7164.
7. Herbert, V. B_{12} Deficiency in AIDS. *JAMA* 1988; 260:2837.
8. Herbert V, Fong W, Gulle V, and Stopler T. Low holotranscobalamin II is the earliest serum marker for subnormal vitamin B_{12} absorption in patients with AIDS. *Am J Hem* 1990; 34(2):132–39.
9. Hansen M, Gimsing P, Ingeberg S, Jans H, and Nexo E, Cobalamin binding proteins in patients with HIV infection. *Eur J Haematol* 1992; 48:228–231.
10. Remacha, A. Acquired Immune Deficiency Syndrome and vitamin B_{12}. *Eur J Haematol* 1989; 42:(5) 506.
11. Remacha A, Riera A, Cadafalch J, and Gimferrer E. Vitamin B_{12} abnormalities in HIV infected patients. *Eur J Haematol* 1991; 47(1):60–4.
12. Baum MW, Beach R, Morgan R, et al. Vitamin B_{12} and cognitive function in HIV infection. San Francisco, Sixth Int Conf on AIDS, June 21–24, 1990, Abstract F. B. 32.
13. Kieburtz KD, Giang DW, Schiffer RB, and Vakil N. Abnormal vitamin B_{12} metabolism in human immunodeficiency virus infection: Association with neurological dysfunction. *Arch Neurol* 1991; 48(3):312–4.
14. Stern RA, Singer NG, Perkins DO, et al. Neurobehavioral impairments in early asymptomatic HIV infection: The effects of education, CD4 count, B_{12} level and depression. Seventh Int Conf on AIDS. Florence, June 16–21, 1991. Abstract TH.B.88.
15. Beach RS, Morgan R, Wilkie F, et al. Plasma vitamin B_{12} level as a potential cofactor in studies of human immunodeficiency virus type 1 — related cognitive changes. *Arch Neurol* 1992; 49:501–506.
16. Malcolm JA, Tynn PF, Sutherland DC, Dobson P, Kelson W, and Carlton J. Trace metal and vitamin deficiencies in AIDS. Sixth Int Conf on AIDS, San Francisco, June 21–24, 1990, Abstract Th. B. 206.
17. Bogden JD, Baker H, Frank O, et al. Micronutrient status and human immunodeficiency virus infection. *Ann NY Acad Sci* 1990; 587:189–95.
18. Butterworth RF, Gaudreau C, Vincelette T, Bourgault AM, Lamotine F, and Nutini A. Thiamine deficiency and Wernicke's encephalopathy in AIDS. *Metab Brain Dis* 1991; 6(4):207–12.
19. Boudes P, Zittoun J, and Sobel A. Folate, Vitamin B_{12} and HIV infection. *Lancet* 1990; 335:1401–2.
20. Zeitz M, Ullrich R, Heise W, Bergs C, L'age M, and Riecken EO. Malabsorption is found in early stages of HIV infection and independent of secondary infection. Seventh Int Conf on AIDS. Florence, June 16–21, 1991. Abstract W.B.90.

21. Livrozet JM, Bourgeay-Causse M, and Fayol V. Vitamin status (Folinic acid and vitamin B_{12}) at the first blood analysis in 140 HIV seropositive patients. Eighth Int Conf on AIDS. Amsterdam, June 19–24, 1992. Abstract PUB 7318.

22. Beach RS, Mantero-Atienza E, Shor-Posner G, et al. Specific nutrient abnormalities in asymptomatic HIV-1 infection. *AIDS* 1992; 6(7):701–708.

23. Herbert V, Jacobson J, Shevchuk O, et al. Vitamin B_{12}, folate and lithium in AIDS. *Clin Res* 1989; 37:594A.

24. Herbert V, Fong W, Jacobson J, et al. Less than 20 pm B_{12} on Transcobalamin II/ml serum predicts inability to absorb B_{12} from food in AIDS patients. *Clin Res* 1989; 37:853A.

25. Herbert V, Jacobson J, Fong W, and Stopler T. Lithium for Zidovudine induced neutropenia in AIDS: In reply. *JAMA* 1989; 262:776.

26. McKinsey D, Durfee D, and Kurtin P. Megaloblastic pancytopenic associated with dapsone and trimethoprim treatment of pneumocystis carinii pneumonia in the acquired immunodeficiency syndrome. *Arch Intern Med* 1989; 149:965.

27. Richman DD, Fischl MA, Greco MH, et al. The toxicity of AZT in the treatment of patients with AIDS and AIDS-related complex. *N Engl J Med* 1987; 317:192–8.

28. Baum MK, Beach R, Mantero-Atienza E, et al: Predictors of change in immune function: longitudinal analysis of nutritional and immune status in early HIV-1 infection. Eighth Int Conf on AIDS. Amsterdam, July 19–24, 1992. Abstract MC 3127.

29. Falutz J, Paltiel O, DiGirolamo A, and Tsoukas CM. Hematologic consequences of low vitamin B_{12} levels in HIV infection. Seventh Int Conf on AIDS. Florence, June 16–21, 1991. Abstract M.B. 2306.

30. Maria-Soledad N, Charekhenian S, Cardon B, and Rozenbaum W. Vitamin B_{12} supplements in patients treated with Zidovudine. Montreal, Fifth Int Conf on AIDS, June 4–9, 1989, Abstract T.B.P. 307.

31. McCutchen JA, Ballard C, Freeman B, Bartok A, and Richman D. Cyanocobalamin (Vitamin B_{12}) supplementation does not prevent the hematologic toxicity of Azidothymidine (AZT). Fifth Int Conf on AIDS, Montreal, June 4–9, 1989, Abstract M.B.P. 325.

32. Clotet B, Gimero JM, Jou A, et al. Toxicity of Zidovudine (AZT) in patients with AIDS. Fifth Int Conf on AIDS, June 4–9, 1989, Abstract T.B.P. 308.

33. Levy RM, Bredeser DE, and Rosenblum ML. Neurologic complication of HIV infection. *Am Fam Physician* 1990; 41:517–536.

34. Mantero-Atienza E, Beach RS, Van Riel F, et al. Low vitamin B_6 levels and immune dysregulation in HIV-1 infection. Fifth Int Conf on AIDS, Montreal, June 4–9, 1989, Abstract Th. B.P. 313.

35. Baum MK, Mantero-Atienza E, Shor-Posner G, et al. Association of vitamin B_6 status with parameters of immune function in early HIV-1 infection. *J Acquir Immune Defic Syndr* 1991; 4(11):1122–1132.

36. Mantero-Atienza E, Baum M, Beach R, Javier J, Morgan R, and Eisdores C. Vitamin B_6 and immune function in HIV infection. Sixth Int Conf on AIDS, San Francisco, June 21–24, 1990, Abstract 3123.

37. Coulston A, McCorkindale C, Dybevik W, and Merrigan T. Nutritional status of HIV + patients during the early stages of the disease. Sixth Int Conf on AIDS, San Francisco, June 21–24, 1990, Abstract Th. B. 200.

38. Miller LT and Kerkvliet NI. Effect of vitamin B_6 on immunocompetence in the elderly. *Ann NY Acad Sci* 1990; 587:49–54.

39. Anonymous. Vitamin B6 and immune function in the elderly and HIV seropositive subjects. *Nutr Rev* 1992; 50(5):145–147.

40. Shor-Posner G, Blaney N, Feaster D, et al. Anxiety and depression in early HIV-1 infection and its association with vitamin B_6 status. Eighth Int Conf on AIDS. Amsterdam, July 19–24, 1992. Abstract POB 3711.

41. Beach R, Mantero-Atienza E, Van Riel F, Morgan R, and Fordyce-Baum MW. Nutritional abnormalities in early HIV-1 infection — plasma vitamin levels. Fifth Int Conf on AIDS, Montreal, June 4–9, 1989, Abstract Th. B.O. 40.

42. Davtyan DG and Vinters HV. Wernicke's encephalopathy in AIDS patients treated with Zidovudine. *Lancet* 1987; 2:919–920.

43. Hutchin KC. Thiamine deficiency, Wernicke's encephalopathy and AIDS. *Lancet* 1987; 2:1200.

44. Foresti V and Connfalonieri F. Wernicke's encephalopathy in AIDS. *Lancet* 1987; 3:1499.

45. Allworth AM and Kemp RJ. A case of acute encephalopathy caused by the human immunodeficiency virus apparently responsive to Zidovudine. *Med J of Australia* 1989; 151:285–6.

46. Javier JJ, Fordyce-Baum MK, Beach RS, Gavancho M, Cabrejos C, and Mantero-Atienza. Antioxidant micronutrients and immune function in HIV-1 infection. *FAJEB* 1990; 4(4):A940 (Abstract).

47. Dushimimana A, Graham NMH, Humphrey JH, et al. Maternal vitamin A levels and HIV related birth outcome in Rwanda. Eighth Int Conf on AIDS. Amsterdam, July 19–24, 1992. Abstract PoC4221.

48. Karter D, Karter AJ, Yarrish R et al. Vitamin A deficiency in patients with AIDS: A cross sectional study. Eighth Int Conf on AIDS. Amsterdam, July 19–24, 1992. Abstract POB 3698.

49. Somner A. Vitamin A status, resistance to infection and childhood mortality. *Ann NY Acad Sci* 1990; 587:17–23.

50. Scrimshaw NS, Taylor CE, and Gordon JE. *Interaction of Nutrition and Infection.* World Health Organization Monograph Series #57, 1968, p. 87–94.

51. Schmidt K. Antioxidant vitamins and B-carotene: Effects on immunocompetence. *Am J Clin Nutr* 1991; 53:3835–3855.

52. Diplock AT. Antioxidant nutrients and disease prevention: An overview. *Am J Clin Nutr* 1991; 53:1895–1935.

53. Schofer H, Ochsendorf FR, Helm EB, and Milbradt R. Treatment of oral hairy leukoplakia in AIDS patients with vitamin A acid (topically) as acyclovir (systemically). *Dermatologica* 1987; 174:150–153.

54. Watson RR, Yahya MD, Darban HR, and Prabhala RH. Enhanced survival by vitamin A supplementation during a retrovirus infection causing murine AIDS. *Life Sciences* 1988; 43:13–18.

55. Fryburg DA, Mark R, Askenase PW, and Patterson TF. The immunostimulatory effects and safety of beta carotene in patients with AIDS. Eighth Int Conf on AIDS. Amsterdam, July 19–24, 1992. Abstract POB 3458.

56. Coodley G. Beta carotene therapy in human immunodeficiency virus infection. *Clin Res* 1991; 39(2):634A (Abstract).

57. Garewal HS, Ampel NM, Watson RR, Prabhala RH, and Dols CL. A preliminary trial of beta carotene in subjects infected with the Human Immunodeficiency virus. *J Nutr* 1992; 122:35:728–732.

58. Coodley GO, Nelson HD, Loveless MO, and Folk C. Beta carotene in HIV infection. *J Acquir Immune Defic Syndr* 1993; 6:272–276.

59. Cathcart RF. Vitamin C in the treatment of acquired immune deficiency syndrome. *Med Hypotheses* 1984; 14:423–433.

60. Harekeh S and Jariwalla RJ. Comparative study of the anti-HIV activities of ascorbate and thol-containing reducing agents in chronically HIV infected cells. *Am J Clin Nutr* 1991; 54(6 suppl):12315–12355.

61. Harekeh S, Jariwalla RJ, and Pauling L. Suppression of human immunodeficiency virus replication by ascorbate in chronically and acutely infected cells. *Proc Natl Acad Sci* 1990; 87(18):7245–9.

62. Jariwalla RJ and Harekeh S. HIV Suppression by ascorbic acid and its enhancement by a glutathione precursor. Eighth Int Conf on AIDS. Amsterdam, July 19–24, 1992. Abstract POB 3697.

63. Locardi C, Petrini C, Boccoli G, et al. Increased human immunodeficiency virus (HIV) expression in chronically infected U937 cells upon in vitro differentiation by hydroxyvitamin D3: roles of interferon and tumor necrosis factor in regulation of HIV production. *J Virol* 1990; 64(12):5874–82.

64. Kitano K, Baldwin GC, Raines MA, and Golde DW. Differentiating agents facilitate infection of myeloid leukemic cell lines by monocytropic HIV-1 Strains. *Blood* 1990; 76(10):1980–8.

65. Skolnick PR, Jahn B, Wang MZ, et al. Enhancement of human immunodeficiency virus replication in monocytes by 1,25 dihydroxycholecalciferol. *Proc Natl Acad Sci* 1991; 88(15):6632–6.

66. Jahn B, Wang M, Griffith A, Krane S, and Skolnick P. 1,25 dihydroxycholecalciferol $(1,25(OH)_{20}O_3$, vitamin D3) and lipopolysaccharide synergistically enhance HIV-1 replication in monocytes. Eighth Int Conference on AIDS. Amsterdam, July 19–24, 1992. Abstract PoA 2478.

67. Connor RI, Rigby WF. 1 alpha, 25-dihydroxyvitamin D3 inhibits productive infection of human monocytes by HIV-1. *Prochem Biophys Res Commun* 1991; 176(2):852–9.

68. Girasole G, Wang JM, Pedrazzoni M, et al. Augmentation of monocyte chemotaxis by 1-alpha-25-dihydroxyvitamin D3 stimulation of defective migration in AIDS patients. *J Immunol* 1990; 545(8):2459–64.

69. Tobler A, Gasson T, Reichel H, Norman AW, and Koeffler HP. Granulocyte-macrophage colony-stimulating factor: sensitive and receptor mediated regulation by 1,25 Dihydroxyvitamin D_3 in normal human peripheral blood lymphocytes. *J Clin Invest* 1987; 74:1700–1705.

70. Hollins TD. T4 cell receptor distortion in acquired immune deficiency syndrome. *Med Hypotheses* 1988; 26:107–111.

71. Kline K, Rao A, Romach E, Kidao S, Morgan TJ, and Sanders BG. Vitamin E effects on retrovirus-induced immune dysfunctions. *Ann NY Acad Sci* 1990; 587:294–296.

72. Odeleye OE and Watson RR. The potential role of vitamin E in the treatment of immunologic abnormalities during acquired immune deficiency syndrome. *Prog Food Nutr Sci* 1991; 15(1–2):1–19.

73. Gogu SR, Beckman BS, Rangan SRS, and Agraral KC. Increased therapeutic efficacy of Zidovudine in combination with vitamin E. *Biochem Biophys Res Commun* 1989; 165, 1:401–407.

74. Qualtiere LF, Zbitnew A, Heise-Qualtiere JM, and Conly J. Menaquinone (bacterial vitamin K) inhibits HIV-1 induced syncytia formation but not HIV-1 replication. Fifth Int Conf on AIDS, June 4–9, 1989, Abstract M.C.P. 147.

75. Herbert V, Jacobson J, Colman N, et al. Negative folate balance in AIDS. *FASEB J* 1989; 3(4):A1278.

76. Revell P, O'Doherty MJ, Tang A, and Savidge GF. Folic acid absorption in patients infected with the human immunodeficiency virus. *J Intern Med* 1991; 230(3):227–31.

77. Tilkian SM and Lefevre G. Altered folate metabolism in early HIV Infection. *JAMA* 1988; 259:3128.

78. Smith I, Howells DW, Kendall B, Levinsky R, and Hyland K. Folate deficiency and demyelination in AIDS. *Lancet* 1987; 3:215.

79. Surtees R, Hyland K, and Smith I. Central nervous system methyl group metabolism in children with neurological complications of HIV infection. *Lancet* 1990; 335:619–621.

80. Blair JA and Heales SRJ. Folate deficiency and demyelination in AIDS. *Lancet* 1987; 3:509.

81. McKinsey DS, Durfee D, and Kurtin PJ. Megaloblastic pancytopenia associated with Dapsone and Trimethoprim treatment of pneumocystis carinii pneumonia in the acquired immunodeficiency syndrome. *Arch Intern Med* 1990; 150:1141.

82. Allegra CJ, Chabner BA, Tuazon CU, et al. Trimetrexate for the treatment of Pneumocystis Carinii Pneumonia in patients with the acquired immunodeficiency syndrome. *N Engl J Med* 1987; 317(16):978–985.

83. Sattler FR, Allegra CJ, Verdegem TD, et al. Trimetrexate–leucovorin dosage evaluation study for treatment of *Pneumocystitis Carinii Pneumonia*. *J Infect Dis* 1990; 161(1):91–6.

84. Hollander H. Leukopenia, trimethoprim-sulfamethoxazole and folinic acid. *Ann Intern Med* 1985; 102:138.

85. Bygbjerg IC, Lund JT, and Harding M. Effect of folic and folinic acid on cytopenia occurring during co-trimoxazole treatment of *Pneumocystitis Carinii Pneumonia*. *Scand J Infect Dis* 1988; 20(6):685–6.

86. Summerbell C, Gazzard B, and Catalan J. The nutritional knowledge, attitudes, beliefs and practices of male HIV positive homosexuals. Seventh Int Conference on AIDS. Florence, June 16–21, 1991. Abstract W.D.4209.

87. Grosvenor M, Tai V, Novak D, et al. Nutritional supplementation by HIV infected persons. Seventh Int Conf on AIDS. Florence, June 16–21, 1991. Abstract M.B.2194.

88. Priestley J. Nutrient replacement therapy enhances survival and laboratory parameters of HIV-positive patients. Eighth Int Conf on AIDS. Amsterdam, July 19–24, 1992. Abstract POB 3710.
89. Baum MK, Shor-Posner G, Bonveh PE, et al. Interim dietary recommendations to maintain adequate blood nutrient levels in early HIV-1 infection. Eighth Int Conf on AIDS. Amsterdam, July 19–24, 1992. Abstract POB 3675.

Chapter

8

The B-Complex Vitamins, Immune Regulation, Cognitive Function, and HIV-1 Infection

Marianna K. Baum

Nancy Cure

and

Gail Shor-Posner

INTRODUCTION

The synergistic interaction of malnutrition and infection has long been recognized.[1] Infectious illness influences nutritional status which, in turn, affects host susceptibility to infection.[2] The interrelationships between infection, nutritional status and immune function are especially apparent in individuals infected with the Human Immunodeficiency Virus (HIV-1), who exhibit impaired immune function and altered nutritional status.

MALNUTRITION AND HIV-1 DISEASE

The immunological consequences and clinical manifestations associated with HIV-1 disease resemble those found in protein-energy malnutrition (PEM).[3-6] Malnutrition is frequently encountered in later stages of HIV-1

0-8493-7842-7/94/$0.00+$.50
© 1994 by CRC Press, Inc.

infection[7] and as reported by Gray,[4] both patients with AIDS and individuals with PEM experience multiple infections of viral, bacterial, parasitic and mycotic origin. The immunological abnormalities observed in malnutrition and HIV-1 infection include decreased T lymphocyte and CD4 cells, reduced secretory IgA, and impaired primary and secondary delayed cutaneous hyper-sensitivity responses.[8-10]

The wasting and malnutrition associated with the advanced stages of HIV-1 infection have been related to a significant alteration in total body potassium, reflecting changes in total lean mass.[11] The degree of lean body mass depletion in the AIDS patient appears to be closely linked to the time interval to mortality.[11] Those patients who suffer from the greatest degree of wasting experience the most rapid rate of mortality.[12] Additional evidence that nutritional status may be an important determinant of survival in AIDS is provided by studies revealing an association between body weight loss, low serum albumin levels and mortality.[11,13] It should be noted that with appropriate treatment of opportunistic infections and adequate nutritional support, patients appear to have better clinical outcomes as well as reduced morbidity and mortality.[14,15]

SPECIFIC NUTRITIONAL ABNORMALITIES AND HIV-1 INFECTION

While nutritional factors have not been postulated to play a primary etiologic role in the development of HIV-1 infection, they appear to be a likely cofactor in the progression of HIV-1 infection and the clinical appearance of AIDS.[5,16,17] In addition to generalized PEM, AIDS patients exhibit deficiencies of a variety of nutrients, including low levels of plasma zinc,[18-20] selenium,[21] folic acid[22] and vitamin B_{12}.[23-25]

Our recent studies, conducted in asymptomatic CDC Stage II,III HIV-1-infected homosexual men, indicate that even prior to the development of significant symptoms, HIV-1-infected patients demonstrate a number of nutritional alterations.[26] Despite dietary intakes that meet or even exceed the recommended dietary allowances,[27] an examination of serum biochemical parameters reveals that up to 67% of the HIV-1-seropositive patients have at least one nutritional deficiency with 36% of the population exhibiting multiple abnormalities.[26]

The wide range of biochemical abnormalities observed during the early stages of HIV-1 infection includes low plasma levels of some of the B vitamins, with either overt or marginal vitamin B_6 deficiency observed in 53% of the subjects and alterations in vitamin B_{12} and riboflavin in approximately 25% of the participants. Plasma values were within the range considered to be normal for thiamin and folate, however, in the majority of the HIV-1-infected men.

VITAMIN B COMPLEX AND IMMUNITY

Considerable evidence, obtained mainly from animal experiments, has accumulated to show an important role for the B vitamins (pyridoxine, pantothenic acid, riboflavin, folate and cobalamin) in immune regulation. Folic acid, for example, has been demonstrated to influence immunological function, and produce a profound myelosuppressive/megaloblastic effect upon bone marrow in deficient animals and patients.[28] In both folate-deficient humans and experimental animals, antibody responses to specific antigens, T cell number, response to mitogens, and multiple abnormalities of neutrophil function have been described.[29] Because of its essential role in the synthesis of thymidylate, vitamin B_{12} appears to be necessary for all proliferating cells[28] and should be assumed to contribute to the support of immune system function and the adequacy of other host defensive mechanisms.

Deficiency of vitamin B_6, which is required for normal nucleic acid and protein synthesis, and cellular multiplication, has been demonstrated to affect immune function, influence tumor susceptibility,[30] and impair dermal response against antigens.[31] Vitamin B_6 deficient rats display a decrease in thymic weight,[32] and reduction in the relative and absolute number of lymphocytes in the peripheral blood as well as the thoracic duct lymph cell count.[33]

Vitamin B_6 deficiency in humans has been linked to profound immunological effects including delayed cutaneous hypersensitivity, altered peripheral white cell count, decreased cell-mediated response, depressed proliferation responses and diminished antibody response to vaccines.[9,31,34-37] A clear vitamin B_6 impact on immune response is apparent in elderly individuals, who exhibit a significant reduction in the percentage and total number of lymphocytes, proliferative responses of peripheral blood lymphocytes to T and B cell mitogens and IL-2 production, with pyridoxine depletion.[38] The majority of these parameters, excluding lymphocyte percentage and number, returned to baseline levels when intake was normalized. Consistent with these findings,[39] Talbot et al. have demonstrated that pyridoxine supplementation improves lymphocyte function in healthy elderly persons, suggesting an important role for vitamin B_6 nutriture in stimulating both T and B cell mitogen response.

VITAMIN B₆, IMMUNE FUNCTION
AND HIV-1 DISEASE

We have recently demonstrated that vitamin B_6 deficiency may contribute to the immunodeficiency observed in the early HIV-1 infected patient.[40] Our investigations, conducted in asymptomatic HIV-1-infected homosexual men with variable degrees of immune dysregulation (mean CD4 cell count 493 ± 282), have revealed a significant relationship between vitamin B_6 status and functional parameters of immunity. Specifically, overtly vitamin B_6-deficient participants exhibited significantly decreased lymphocyte responsiveness to the mitogens phytohemagglutinin and pokeweed, and reduced natural killer cell cytotoxicity, as compared to individuals with adequate vitamin B_6 status. No association was noted, however, between vitamin B_6 levels and the T lymphocyte subpopulation or serum immunoglobulin levels.

The basic etiology of vitamin B_6 deficiency in HIV-1-infected individuals remains unknown at present, but appears to be multifactorial in nature. HIV-1-infected patients may have an enteropathy related to the HIV infection per se, which, in addition to the nutritional problems associated with the presence of opportunistic infections and malignancies, could result in a considerable degree of malabsorption.[41,42] Renal complications frequently occur in HIV-1 disease[43] possibly resulting in increased renal losses of vitamin B_6. Although vitamin B_6 deficiency may involve increased hepatic metabolism, such as that which occurs in patients with a history of chronic ethanol abuse,[44] our study did not reveal any correlation between history of liver disease, or liver function test and vitamin B_6 deficiency in HIV-1-infected individuals.

VITAMIN B_{12} STATUS, IMMUNE FUNCTION AND HIV-1 DISEASE

The importance of vitamin B_{12} in immune function has been elucidated from studies in patients with untreated primary pernicious anemia. Whereas unusually high amounts of vitamin B_{12} exposure have been associated with an increased lymphocyte response to phytohemagglutinin stimulation in animal studies,[45] diminished uptake of [3H] thymidine by lymphocytes has been reported in patients with pernicious anemia.[46] Supplementation of anemic patients with vitamin B_{12} has been shown to reverse CD4/CD8 imbalance, suggesting immunodysregulation secondary to vitamin B_{12} deficiency.[47,48] Deficiency of vitamin B_{12} appears to impair the metabolic activities associated with phagocytosis[49] producing a marked decrease in phagocytosis-associated activation of the hexose monophosphate shunt, along with a slight to moderate impairment of intraleukocytic microbicidal activity.[50] These abnormalities of function are reversed after vitamin B_{12} replacement,[50] supporting an important role for vitamin B_{12} in normal cell metabolism and function.

Low levels of serum vitamin B_{12} have been reported during the early stages of HIV-1 disease,[26,51] with overt deficiency (<200 pg/mL) present in 12% of the seropositive subjects and marginal deficiency (200–240 pg/mL), presenting in an additional 11% of the participants. The low levels of vitamin B_{12} were observed despite adequate dietary intake of vitamin B_{12}[27] and the lack of classical indicators of megaloblastic anemia. Our longitudinal analyses of nutritional status over time indicated that change in plasma vitamin B_{12} level, from biochemically deficient to adequate, was associated with an improvement in markers of disease progression (CD4 cell count and AIDS index, a composite measurement of disease progression).[52,53] In other studies, HIV-1-infected individuals with low serum vitamin B_{12} levels have been documented to have lower hemoglobin, leukocytes, lymphocytes, CD4 lymphocytes and CD4/CD8 lymphocyte ratio, as well as higher mortality, relative to HIV-1 seropositive patients with physiological serum vitamin B_{12} levels.[54] None of the individuals exhibited megaloblastic changes in their peripheral blood smears and screening for gastric and intestinal dysfunction produced negative results in nearly all patients tested.

The deficiency of vitamin B_{12} observed in HIV-1-seropositive individuals may be due, in part, to decreased vitamin B_{12} absorption produced by the effect of HIV on the gastric mucosa and/or the presence of opportunistic infections in the small bowel. Change of the gastro-intestinal pH and alterations in the intrinsic factor also contribute to vitamin B_{12} malabsorption.[55] Abnormal Schilling tests in association with histologic evidence of chronic inflammation in patients without diarrhea or weight loss, suggest vitamin B_{12} malabsorption occurs early in the course of the HIV infection. The malabsorption may be related to localization of HIV in cellular reservoirs of the terminal ileum.[56]

The decreased levels of vitamin B_{12} may also be caused by changes in the concentration of cobalamin-binding proteins that carry cobalamin in plasma. Hansen et al[57] have demonstrated low concentrations of cobalamin-saturated binding proteins (holo-transcobalamin and holo-haptocorrin), and increased concentrations of cobalamin-unsaturated binding proteins (apo-trancobalamin, apo-haptocorrin) in HIV-1-infected subjects, suggesting that low plasma cobalamin may not reflect a low concentration of transcobalamin or haptocorrin. It has been hypothesized[55] that DNA-synthesizing cells of the hematopoietic, immunologic, and neurologic systems have surface receptors solely for holo-transcobalamin II, the earliest serum marker of subnormal vitamin B_{12} absorption. Cells containing these surface receptors rapidly become dysfunctional, due to vitamin B_{12} deficiency when holo-transcobalamin II is low, while cells which also have surface receptors for holo-haptocorrin remain vitamin B_{12} replete. Recent data suggest that AIDS may prevent vitamin B_{12} deficiency rise in serum homocysteine.[58]

The investigations briefly summarized above demonstrate that in HIV-1 infection, vitamin B_6 and vitamin B_{12} deficiency are relatively widespread and potentially important cofactors of HIV-1 disease progression. Development of such deficiencies may significantly influence the course of HIV-1 infection, underscoring the need for determining nutritional status during the early stages of HIV disease. Further investigation will be required to determine whether normalization of nutrient status would prove beneficial in maintaining optimal immune function.

VITAMIN B COMPLEX, NEUROPSYCHOLOGICAL FUNCTION AND HIV-1 DISEASE

In addition to its profound effect on immune regulation, nutritional status has been demonstrated to play an important role in central nervous system function,[59] with nutrient deficiencies of thiamin, pyridoxine, niacin, pantothenic acid, biotin, vitamin B_{12} and folate frequently associated with psychoneurological symptoms ranging from peripheral neuropathies to spinal cord degeneration and global cognitive impairment.[59] Damage to the central nervous system in HIV-1 infection is manifested by many neurological symptoms similar to those associated with nutritional deficiencies, including diffuse and regional encephalopathies, myelopathy, meningitis, intraaxial cranial neuropathies, and retinopathy.[60-62]

Our research studies also suggest the possibility that inadequate vitamin B_6 status may contribute to the progression of neuropsychological abnormalities associated with HIV-1 infection. Specifically, our investigations have demonstrated significant associations between vitamin B_6 status and measures of cognition involving reaction time in HIV-1 infected subjects.[63] In particular, a decline in vitamin B_6 status over time was significantly associated with deterioration of performance in tasks related to optimal peripheral psychomotor function. This decrease in reaction time was in contrast to the improvement observed when vitamin B_6 status remained adequate.[63]

Deficiency of vitamin B_{12} has also been linked to a wide variety of neurological, psychiatric, and cognitive alterations in a number of clinical settings.[64-67] Peripheral neuropathy and myelopathy, which have been associated with vitamin B_{12} deficiency,[68] have also been noted in HIV-1 disease.[69-72] The prevalence of abnormal vitamin B_{12} metabolism in HIV-1-infected patients referred for neurological evaluation has been especially evident in individuals with concomitant myelopathy and neuropathy,[73] which are the most commonly encountered in vitamin B_{12} deficiency due to pernicious anemia.[57]

Our studies[74] have revealed a strong relationship between vitamin B_{12} status and cognitive function during the early stages of HIV-1 infection, particularly in areas of information processing speed and visuospatial problem-solving abilities. Subjects with low plasma vitamin B_{12} levels score more poorly than participants with normal levels of vitamin B_{12} on the Posner Letter Matching, Figure Visual Scanning and Discrimination of Pictures time tasks, regardless of their HIV-1 status. These results suggest that vitamin B_{12} deficiency could be a contributing factor to slower information processing speeds in asymptomatic HIV-1-infected subjects.

SUMMARY

Multiple nutritional abnormalities have been demonstrated to occur relatively early during the course of HIV-1 disease. Deficiencies of vitamin B complex, which are particularly widespread, may contribute to the immune dysregulation and be related to the cognitive impairment frequently associated with HIV-1 infection. Clinical trials are needed to confirm that restoration of adequate plasma vitamin B complex levels improves immune function and cognitive processing in HIV-1 disease. Through the development of appropriate nutritional intervention strategies, it may be possible to improve the quality of life and slow disease progression in the HIV-1-infected individual.

REFERENCES

1. Scrimshaw, N., Taylor, and Gordon, J.S. Interactions of nutrition and infection. *Am J Med Sci* 237:367, 1959.
2. Beisel, W. Single nutrients and immunity. *Am J Clin Nutr* 35(Supp)417, 1982.
3. Beach, R.S. and Laura, P.S. Nutrition and the acquired immunodeficiency syndrome. *Ann Intern Med* 99:565, 1985.
4. Gray, R.H. Similarities between AIDS and PCM. *Am J Publ Health* 73:1332, 1983.
5. Jain, V.K. and Chandra, R.K. Does nutritional deficiency predispose to the acquired immune deficiency syndrome? *Nutr Res* 4:537, 1984.
6. Lesbourdes, J.L., Chassignol, S., Ray, E., et al. Malnutrition and HIV infection in children in the central African republic. *Lancet* 2:337, 1986.
7. Excler, J.L., Standaert, B., Ngendandumwe, E., and Piot, P. Malnutrition et infection a HIV chez l'enfanten milieu hospitalier au Burundi. *Pédiatrie* 42:715, 1987.
8. Neumann, C.G., Lawlor, G.J., Jr., Strehm, E.R., et al. Immunologic responses in malnourished children. *Am J Clin Nutr* 28:89, 1975.
9. Cunningham-Rundles, S. Effects of nutritional status on immunological functions. *Am J Clin Nutr* 35:102, 1982.
10. Fauci, A.S. The human immunodeficiency virus: ineffectivity and mechanisms of pathogenesis. *Science* 239:717, 1988.
11. Kotler, D.P., Wang, J., and Pierson, Jr, R.N. Body composition studies in patients with the acquired immunodeficiency syndrome. *Am J Clin Nutr* 42:1255, 1985.
12. Kotler, D.P., Tierney, A.R., Wang, J., and Pierson, Jr, R.N. Magnitude of body cell mass depletion and the timing of death from wasting in AIDS. *Am J Clin Nutr* 50:444, 1989.

13. Cheblowski, R.J., Grosvenor, M.B., Bernhard, N.H., Morales, L.S., and Bulcarage, L.M. Nutritional status, gastrointestinal dysfunction and survival in patients with AIDS. *Am J Gastroenterol* 84:1288, 1989.

14. Kotler, D.P., Tierney, A.R., Altilio, D., Wang, J., and Pierson, Jr, R.N. Body mass repletion during gancilovier therapy of cytomegalovirus infections in patients with the acquired immunodeficiency syndrome. *Arch Intern Med* 149:901, 1989.

15. Kotler, D.P., Tierney, A.R., Brenner, S.K., Couture, S., Wang, J., and Pierson, Jr, R.N. Preservation of short-term energy balance in clinically stable patients with AIDS. *Am J Clin Nutr* 51:7–13, 1990.

16. Beach, R.S., Cabrejos, C., Shor-Posner, G., et al. Nutritional aspects of early HIV infection. In *Nutrition and Immunology,* Rajit Chandra, ed., Biomedical Publisher and Distributors, 1992.

17. Moseson, M., Zeleniuch-Jacquotte, P., Belsito, D.V., et al. The potential role of nutritional factors in the induction of immunologic abnormalities in HIV-positive homosexual men. *J AIDS,* 2:235, 1989.

18. Falutz, J., Tsoukas, C., and Gold, P. Zinc as a cofactor in human immunodeficiency virus-induced immunosuppression. *JAMA* 259:2850, 1988.

19. Fabris, N., Mocchegiani, E., Galli, M., and Lazzarin, A. AIDS, zinc deficiency, and thymic hormone failure. *JAMA* 259:839, 1988.

20. Shoemaker, J.D., Millard, H.C., and Johnson, P.B. Zinc in human immunodeficiency virus infection. *JAMA* 260:1881, 1988.

21. Dworkin, B.M., Rosenthal, W.S., Wormser, G.P., and Weiss, L. Selenium deficiency in acquired immunodeficiency syndrome. *JPEN* 10:405–407, 1986.

22. Smith, J., Howells, D.W., Kendall, B., Lavensky, R., and Hyland, K. Folate deficiency and demyelination in AIDS. *Lancet* 2:25, 1987.

23. Burkes, R.L., Cohen, H., Krailo, M., Sinow, R.M., and Carmel, R. Low serum cobalamin levels occur frequently in the acquired immunodeficiency syndrome and related disorders. *Eur J Haematol* 38:141, 1987.

24. Harriman, G.R., Smith, P.D., Horne, M.K., Fox, C.H., et al. Vitamin B_{12} malabsorption in patients with acquired immune deficiency syndrome. *Arch Intern Med* 149:2039, 1989.

25. Herbert, V. B_{12} deficiency in AIDS. *JAMA* 260:2837, 1988.

26. Beach, R.S., Mantero-Atienza, E., Shor-Posner, G., et al. Specific nutrient abnormalities in asymptomatic HIV-infection. *AIDS* 6:701, 1992.

27. Baum, M.K., Shor-Posner, G., Bonvehi, P., et al. Influence of HIV infection on vitamin status and requirements. *Ann NY Acad Sci* 165, 1992.

28. Herbert, V. Biology of disease: the megaloblastic anemias. *Lab Invest* 52:3–19, 1985.

29. Gershwin, M.E., Beach, R.S., and Hurley, L.S. *Nutrition and Immunity.* Academic Press, Orlando, FL 1985.

30. Ha, C., Kerkvliet, N.I., and Miller, L.T. The effect of vitamin B_6 deficiency on host susceptibility to Maloney Sarcoma Virus-induced tumor growth in mice. *J Nutr* 114:9338, 1984.

31. Axelrod, A.E., Trakatellis, A.C., Bloch, H. et al. Effect of pyridoxine deficiency upon delayed hypersensitivity in guinea pigs. *J Nutr* 79:161, 1963.

32. Moon, W.Y. and Kirksey, A. Cellular growth during prenatal and early postnatal periods in progeny of pyridoxine-deficient rats. *J Nutr* 102:123, 1973.

33. Robson, L.C. and Schwarz, M.R. Vitamin B_6 deficiency and the lymphoid system I. *Cell Immunol* 16:135, 1975.
34. Panush, R. and Delafuente, J. Vitamins and immunocompetence. *Wld Rev Nutr Diet* 45:97, 1985.
35. Hodges, R.E, Bean, W.B., Ohlson, M.A, and Bleiler, R.E. Factors affecting human antibody response IV vitamin B_6 deficiency. *Am J Clin Nutr* 11:180, 1962.
36. Chesloc, K.E. and McCully, M.T. Response of human beings to a low vitamin B_6 diet. *J Nutr* 70:507, 1960.
37. Robson, S. and Schwarz, M.R. Vitamin B_6 deficiency and the lymphoid system. I. Effects on cellular immunity and in vitro incorporation of 3H-uridine by small lymphocytes. *Cell Immunol* 16:135, 1975.
38. Meydani, S.N., Ribaya-Mercado, J.D., Russel, R.M., et al. The effect of vitamin B_6 on the immune response of healthy elderly. In Micronutrients and immune function. *Ann NY Acad Sci,* Adriane Bendich and Ranjit Chandra eds. 587:303, 1990.
39. Talbot, M.C., Miller, L.T., and Kerkvliet, N.I. Pyridoxine supplementation: effect on lymphocyte responses in elderly persons. *Am J Clin Nutr* 46:659, 1987.
40. Baum, M.K., Mantero-Atienza, E., Shor-Posner, G., et al. Association of vitamin B_6 status with parameters of immune function in early HIV-1 infection. *J AIDS* 4:1122, 1992.
41. Guillin, J.S., Shike, M., Alcock, N., et al. Malabsorption and mucosal abnormalities of the small intestine in the acquired immunodeficiency syndrome. *Ann Intern Med* 102:619, 1985.
42. Sewenkabs, N., Mugerwa, R.D., Goodgame, R., et al. Enteropathic AIDS in Uganda. An endoscopic, histological and microbiological study. *AIDS* 1:9, 1987.
43. Pardo, V., Aldana, M., Colton, R.M., et al. Glomerular lesions in the acquired immunodeficiency syndrome. *Ann Intern Med* 101:429, 1984.
44. Mitchell, D., Wagner, L., Stone, W.J., et al. Abnormal regulation of plasma pyridoxal 5'-phosphate in patients with liver disease. *Gastroenterology* 7:1043, 1976.
45. Chandra, R.K. and Newberne, P.M. *Nutrition, Immunity and Infection.* Plenum Press, New York, 1979.
46. MacCuish, A.C., Urbaniek, S.J., Goldstone, A.H., et al. PHA responsiveness and subpopulation of circulating lymphocytes in pernicious anemia. *Blood* 44:849, 1974.
47. Kubota, K., Arai, T., Tamura, J., Shirakura, T., and Mouta, T. Restoration of decreased suppressor cells by vitamin B_{12} therapy in a patient with pernicious anemia. *Am J Hematol* 24:221, 1987.
48. Kubota, K., Kurabayaski, H., Kawada, E., Okamoto, K., and Shirakura T. Restoration of abnormally high CD4/CD8 ratio and low natural killer cell activity by vitamin B_{12} therapy in a patient with post gastrectomy megaloblastic anemia. *Int Med* 31(1): 125, 1992.
49. Sbarra, A.J., Selvaraj, R.J., Paul, B., Strauss, R.R., Jacobs, A.A., and Mitchell, Jr, G.W. Bactericidal activities of phagocytes in health and disease. *Am J Clin Nutr* 27:629, 1974.
50. Kaplan, S.S. and Basford, R.E. Effect of vitamin B_{12} and folic acid deficiencies on neutrophil function. *Blood* 47:801, 1976.

51. Mantero-Atienza, E. Baum, M.K., Morgan, R., et al. Vitamin B_{12} in early human immunodeficiency virus-1 infection. *Arch Intern Med* 151:1019, 1991.

52. Baum, M.K., Beach, R.S., Mantero-Atienza, E., et al. Predictors of change in immune functions: Longitudinal analysis of nutritional status in early HIV-1 infection. VII Int Conf on AIDS, MC 3127,329, Florence, 1991.

53. Fahey, J.L., Taylor, M.G., Detels, R., et al. The prognostic value of cellular and serological markers in infection with human immunodeficiency virus type-1. *N Engl J Med* 322:116, 1990.

54. Remacha, A.F., Riera, A., Cadafalch, J., et al. Vitamin B_{12} abnormalities in HIV-infected patients. *Eur J Haematol* 47:60, 1991.

55. Herbert, V., Fong, W., Gulle, V., et al. Low holo- transcobalamin II is the earliest serum marker for subnormal vitamin B_{12} (cobalamin) absorption in patients with AIDS. *Am J Hematol* 34:132, 1990.

56. Harriman, G.R., Smith, P.D., Horne, Mc., et al., Vitamin B_{12} malabsorption in patients with acquired immunodeficiency syndrome. *Arch Intern Med* 149:2039, 1989.

57. Hansen, A., Gimsing, P., Ingeberg, S., et al. Cobalamin binding proteins in patients with HIV infection. *Eur J Haematol* 48:228, 1992.

58. Jacobson, D.W., Green, R., Herbert, V., Longworth, D.L., and Rehm, S. Decreased serum gluthione with normal cysteine and homocysteine levels in patients with AIDS. *Clin Res* 38:56A, 1990.

59. Dreyfus, P.M. Diet and nutrition in neurologic disorders. In: *Modern Nutrition in Health and Disease.* M.E. Shils and V.R. Yound (Eds). Lea & Febiger, Philadelphia, 1458, 1988.

60. Price, R.W., Brew, B., Sidtes, J., Rosenblum, M., Scheck, A.C., and Cleary, P. The brain in AIDS: Central nervous system HIV-1 infection and AIDS dementia complex. *Science* 239:586, 1988.

61. Levy, R.M. and Bredesen, D.E. Central nervous system dysfunction in acquired immunodeficiency syndrome. *J AIDS* 1:41, 1988.

62. Navia, B.A., Jordan, B.D., and Price, R.W. The AIDS dementia complex: I. Clinical features. *Ann Neurol* 19:517, 1986.

63. Wilkie, F., Shor-Posner, G., Mantero-Atienza, E., et al., Association of vitamin B_6 status and reaction time in early HIV-1 infection. Neuroscience of HIV infection. Int Conf on Neuroscience of HIV Infection. Padova, Italy, Abst. 67, 1991.

64. Elsberg, L., Hansen, T., and Rafaelsen, O.J. Vitamin B_{12} concentrations in psychiatric patients. *Acta Psychiatr Scand* 59:145, 1979.

65. Hector, M. and Burton, J.R. What are the psychiatric manifestations of B_{12} deficiency? *J Am Geriatr Soc* 36:1105, 1988.

66. Carmel, R., Karnaze, D.S., and Weiner, J.M. Neurologic abnormalities in cobalamin deficiency are associated with higher cobalamin "analogue" values than are hematologic abnormalities. *J Lab Clin Med* 111:57, 1983.

67. Lindenbaum, J., Healton, E.B., Savage, D.F., et al. Neuropsychiatric disorders caused by cobalamin deficiency in the absence of anemia or macrocytosis. *N Engl J Med* 318:1720, 1988.

68. Former, T. Neurologic complications of vitamin and mineral disorders. In: *Clinical Neurology,* R.J. Joynt, Ed., J.B. Lippincott, Philadelphia, PA, 1989.

69. Bailey, R.O., Boltich, A.L., Venkatesh, et al. Sensory motor neuropathy associated with AIDS. *Neurology,* 38:886, 1988.
70. Cornblath, D.R. and McArthur, J.L. Predominantly sensory neuropathy in patients with AIDS and ARC. *Neurology* 38:794, 1988.
71. Parry, G. Peripheral neuropathies associated with HIV-1 infection. *Ann Neurol* 23(supp):S49, 1988.
72. Petito, C.K. Navia, R.A., Cho, E.S., et al. Vacuolar myelopathy pathologically resembling subacute combined degeneration in patients with the acquired immunodeficiency syndrome. *N Engl J Med* 312:874, 1985.
73. Kieburtz, K.D., Gianz, D.W., Schiffer, R.B., et al. Abnormal vitamin B_{12} metabolism in human immunodeficiency virus infection. *Arch Neurol* 48:312, 1991.
74. Beach, R.S., Morgan, R., Wilkie, F., et al. Plasma vitamin B_{12} level as a potential cofactor in studies of human immunodeficiency virus type 1-related cognitive changes. *Arch Neurol* 49:501, 1992.

Ascorbic Acid and AIDS: Strategic Functions and Therapeutic Possibilities

Raxit J. Jariwalla

and

Steve Harakeh

INTRODUCTION

Ascorbic acid (ascorbate, vitamin C) is an essential nutrient that exhibits multiple physiologic and pharmacologic functions dependent on its concentration in tissues, blood and body fluids. Important metabolic functions of ascorbate include stimulation of collagen and carnitine synthesis, reduction of molecular oxygen during respiratory burst in phagocytic cells and neutralization of toxic free radicals. At higher doses, ascorbate has desirable pharmacologic effects that may have therapeutic value in controlling the progression of HIV infection and cellular damage associated with AIDS. Of primary strategic importance are its antiviral and antioxidant properties, especially its ability to function as an essential antioxidant under conditions of glutathione deficiency. Additionally, ascorbate exhibits antimicrobial activity against a broad range of pathogenic bacteria and plays a significant role in enhancement of cell-mediated immune responses. These functions, along with stimulation of collagen/carnitine synthesis may aid in the prevention/alleviation of opportunistic infections and wasting syndrome associated with advanced stages of AIDS. Here we discuss the physiologic and pharmacologic functions of ascorbate and the therapeutic possibilities they offer for HIV infection and AIDS.

0-8493-7842-7/94/$0.00+$.50
© 1994 by CRC Press, Inc.

METABOLIC FUNCTIONS OF ASCORBATE

Ascorbic acid (vitamin C) is an essential water-soluble nutrient that has to be obtained from an exogenous source because primates and humans lack a key enzyme required for its synthesis.[1] It is usually found in citrus fruits, cruciferous vegetables, peppers and potatoes. As a vitamin, it is required for important metabolic functions, which include hydroxylation of proline and lysine in the synthesis of collagen; synthesis of carnitine required for fatty acid transport and energy production in the mitochondria; reduction of molecular oxygen in the respiratory burst of phagocytes; and as an antioxidant for neutralizing toxic free radicals. Small amounts of vitamin C that are consumed by healthy people are resupplied from food or regenerated in the body from reduction of its oxidized form (dehydroascorbate, DHA) back to ascorbic acid by reducing equivalents (*i.e.,* hydrogen atoms) provided by glutathione (GSH).[2-4] Lack of vitamin C in tissues leads to scurvy, an often fatal condition characterized by structural debilitation of connective tissues affecting bones, cartilage, muscles, and gums, and impairment of wound healing.[5] Scurvy also causes fatigue, weight loss, joint and muscle pain and skin lesions. Abnormally low levels of ascorbic acid produce a scorbutic-like state (hypoascorbemia) or subclinical scurvy compatible with prolonged survival.[6,7]

At high doses, ascorbate has a unique pharmacologic function, exhibiting the ability to serve as an essential antioxidant and primary source of electrons under conditions of drug-induced GSH deficiency or severe free-radical toxicity.[4,8,9] Because antioxidant depletion and a chronic scorbutic-like state are associated with AIDS, metabolic functions of ascorbate essential to preventing these conditions in healthy individuals are potentially relevant to the control and treatment of such sequelae in persons with HIV infection and AIDS. Here, we briefly review those metabolic functions of ascorbate relevant to HIV/AIDS. Extensive general discussion of biochemical functions of ascorbic acid was previously presented in a review by Englard and Seifter.[10] Chemistry of ascorbic acid and the intercellular matrix of special relevance to cancer was previously reviewed by Cameron et al.[11]

Biosynthesis of Collagen and Carnitine

Vitamin C deficiency in scurvy affects connective tissues, primarily as a result of the inability of the organism to synthesize normal collagen fibers.[5] Earlier studies showed that guinea pigs fed a vitamin C-free diet produced less insoluble collagen than the control group receiving an adequate diet.[12-14] Accumulation of a collagen precursor which had a lower amount of hydroxyproline was noted. A significant proportion of the native collagen fiber is composed of the unique amino acid derivatives, hydroxyproline and hydroxylysine. Hydroxyproline stabilizes the collagen structure and induces the secretion of procollagen. Stetten[13] reported that the hydroxylation of proline and lysine occurred following their incorporation into the backbone of the collagen molecule. He found that L-ascorbic acid was required as a reducing agent for hydroxylation of these amino acid residues. Gould and Woessner[14] showed that dietary depletion of ascorbic acid led to the total inhibition of *de novo* hydroxyproline formation at the site of injury induced in skin.

The mechanism by which vitamin C mediates hydroxylation of proline has been studied by several investigators using fibroblasts *in vitro*. Cardinale[15] reported that vitamin C enhanced the activity of the enzyme, prolyl hydroxylase, in cultured mouse fibroblasts. Stassen et al.[16] noted a 5-fold increase in activity of the same enzyme after the addition of ascorbate to the fibroblast growth medium. Ascorbate depletion led to the impairment of hydroxylproline formation and resulted in the generation of a non-fibrous, high molecular weight precursor of collagen. This effect on hydroxyproline formation was identified as the earliest biochemical defect resulting from ascorbate depletion.

Subsequent studies have implied an additional role for ascorbic acid in the regulation of the enzyme activity that catalyzes proline hydroxylation. Levene and Bates[17] observed that the magnitude of proline hydroxylation stimulation by ascorbate in mouse fibroblasts was similar to that occurring in high density cultures in the absence of ascorbate. They suggested that ascorbic acid was involved in the activation of prolyl hydroxylase by causing aggregation of inactive enzyme subunits. This explanation was consistent with the earlier observation of Stassen et al.[16] who had reported that even in the presence of protein synthesis inhibitors, prolyl hydroxylase activity could be significantly increased by addition of ascorbate to the culture medium. McGee and Udenfriend[18] were further able to isolate a protein, immunologically identical to prolyl-hydroxylase, that was enzymatically inactive and had one-third the molecular weight of the inactive enzyme, indicating the existence of a subunit.

Other reports have indicated a more indirect role for ascorbate in collagen synthesis. Thus, Peterkofksy[19] noted that vitamin C deficiency was responsible for the inhibition in guinea pigs in the synthesis of both proteoglycan and collagen. These inhibitory effects were correlated with weight loss during scurvy. However, the inhibition of both functions in cultured cells was reversed with insulin-like growth factor (IGF)-1. It was suggested that vitamin C deficiency induced two IGF-binding proteins responsible for inhibition of collagen and proteoglycan synthesis in scorbutic guinea pigs. Additionally, Houglum et al.[20] found that low ascorbic acid concentrations stimulated collagen biosynthesis independently of hydroxylation. They ascribed this stimulation to a prooxidant effect, possibly related to the induction of lipid peroxidation and reactive aldehyde generated as a result of ascorbic acid oxidation. Finally, it has been suggested that the role of ascorbate in collagen biosynthesis may be related to the restoration of the reduced state of iron (i.e., ferrous ion, Fe^{+2})

required for the activity of key enzymes involved in collagen biosynthesis.[10] Whatever the molecular mechanism of ascorbate action, there is very little doubt that the vitamin plays an essential role in the synthesis and formation of native collagen molecules.

Early features of scurvy, comprising weakness and fatigue, have also been attributed to deficiency in carnitine,[21,22] a quaternary amino acid, involved in the transport of long-chain fatty acids across the mitochondrial membrane. The fatty acids are used for production of energy in the muscles.[22] Ascorbate has been shown to be an essential cofactor in two hydroxylation steps in the synthesis of carnitine from dietary lysine and methionine.[22] Partial deprivation of dietary ascorbate in guinea pigs was shown to result in reduced muscle carnitine.[22] The muscular weakness, lassitude, and apathy in these animals was ascribed to their reduced ability to utilize fatty acids for energy production.[22] Reduced ascorbate levels in serum and tissues of cancer patients have been linked to cachexia, a condition characterized by progressive lassitude and muscular weakness with gradual loss of mean muscle mass.[23,24] Administration of supplemental ascorbate to such patients was reported to result in improvement in vigor, muscle strength, and overall well-being.[25] Although AIDS patients exhibit low ascorbate in serum,[26] the status of carnitine in HIV/AIDS has not been investigated. Since AIDS patients exhibit a clinical picture (wasting syndrome) similar to advanced cancer patients, studies of carnitine status in HIV-infected persons may provide useful information.

Antioxidant Or Free-Radical Scavenging Function

The ability of ascorbic acid to readily donate electrons endows it with the chemical capacity to reduce oxidizing substances or free radicals. Each molecule of ascorbic acid ($C_6H_6O_6H_2$) has two redox-active hydrogen atoms (hydride anions) in its molecular structure that carry two high-energy electrons. The latter are donated to neutralize free radicals. Depending on its concentration, ascorbate can act as either a carrier or a source of high-energy electrons, distinct functions that make it a premier antioxidant in the body.[2,9]

At ordinary concentrations at which it is normally present, ascorbate functions as an electron-carrier akin to the other non-enzymatic radical scavengers such as α-tocopherol (vitamin E), beta-carotene, glutathione (GSH), and the thiol-containing amino acid, cysteine. In an *in vitro* study analyzing the antioxidant activities of various components of human blood plasma, Frei, Ames and co-workers showed that ascorbate was by far the most effective antioxidant that completely protected plasma lipids from oxidative damage caused by peroxyl radicals.[27]

In the role of electron-carrier, the hydride anions of ascorbate are utilized to neutralize free radicals, and, in the process, ascorbate is oxidized to DHA. In well-nourished individuals, ascorbate is regenerated from DHA through its reduction by GSH via the action of glutathione peroxidase.[2-4] Oxidized GSH disulphide produced during this process is reduced back to GSH by $NADPH_2$ (or $FADH_2$) generated during oxidation of carbohydrates and fats.[2,9] This recycling of antioxidants works well as long as the supply of reducing equivalents from the glycolytic and citric acid cycles is maintained to quench free radicals produced in the body.

During inflammation and illness, when the free-radical load in the body exceeds the supply of available reducing equivalents, the oxidized forms of the radical scavengers remain unreduced. In the case of ascorbic acid, DHA gets degraded, as sufficient GSH is not available to reconvert DHA back to ascorbate. This condition of GSH deficiency may lead to hypoascorbemia, predisposing individuals to oxidative cellular damage caused by unquenched free radicals. AIDS patients manifest striking GSH deficiencies and often exhibit symptoms of acute-induced scurvy characterized by life-threatening weight loss, brittle bones and swollen glands.[9]

Acute-induced scurvy is indicative of a high requirement for exogenous ascorbate to restore normalcy. It has been pointed out that at high concentrations (obtained through supplementation), ascorbate can act directly to scavenge free radicals as well as convert oxidized forms of non-enzymatic scavengers (tocopherol, GSH disulfide) to their reduced states.[4,9] Under these conditions, ascorbate functions as a direct source of high-energy electrons,[9] spares GSH[4] and acts as an essential antioxidant in the presence of GSH deficiency.[4]

ASCORBATE EFFECTS ON THE IMMUNE SYSTEM

Vitamin C plays an important role in the regulation and function of the immune system. While it has been shown to stimulate variably serum antibody levels in human volunteers ingesting one or more grams of ascorbate for several days or weeks,[28,29] its most striking and reproducible effects are on the production and proper functioning of the white blood cells that play important roles in cell-mediated immunity.

Lymphoid cells have been shown to accumulate ascorbate to high levels. Existence of a concentration gradient across the leukocyte membrane was suggested from the study of Evans et al.[30] who reported that mononuclear leukocytes had 80 times the level of ascorbic acid than that in blood plasma. Similar observations were made by Bergsten et al.[31] who found millimolar concentrations of ascorbate in purified mononuclear leukocytes. Furthermore, Washko et al.[32,33] demonstrated that in addition to mononuclear cells, polymorphonuclear leukocytes (PMN), such as neutrophils, accumulate ascorbate to millimolar amounts. High intracellular ascorbate concentrations suggest a compelling role for ascorbate in immune functions. Many independent studies have demonstrated the modulation by ascorbate of several aspects of the immune system, including protection from anaphylactic shock, enhancement of delayed hypersensitivity and graft rejection, enhanced phagocytic motility and function, increased T cell responsiveness to mitogens and increased production of interferon.

Effect on Immediate and Delayed Hypersensitivity and Allograft Rejection

Many earlier reports showed beneficial effects of vitamin C in protecting against anaphylactic shock. However, like the studies on antibody production, these observations were ambiguous and poorly reproducible. Feigen et al.[34] pointed out that pervasive problems in older studies were the lack of standardized immune reagent ratios employed and the rapid disappearance of injected vitamin C. By specifying the load of immune reactants and carrying out the immunological reaction in the presence of a constant level of circulating ascorbate, these authors were able to demonstrate enhancement of both antibody production and protection against systemic anaphylaxis by large doses of vitamin C.

There is reproducible evidence in the literature on the effect of vitamin C upon the underlying immune response mediating delayed hypersensitivity and allograft skin reaction. Guinea pigs administered large doses of vitamin C showed an enhanced delayed hypersensitivity skin reaction.[35-37] It was also reported that guinea pigs on a low vitamin C diet accepted skin grafts from other guinea pigs, while those on a high vitamin C diet rejected such grafts.[38]

Effects on Phagocytic Function

Macrophages and neutrophils play important roles in defending the body against infection. These cells have the ability to engulf and destroy foreign bodies, including bacteria and viruses, through phagocytosis. It was first demonstrated by Cottingham and Mills[39] that the presence of ascorbic acid was important in maintaining phagocytic activity of the white blood cells. A decrease in phagocytic activity was evident in the presence of ascorbic acid deficiency. These findings were later confirmed by DeChatelet et. al.[40]

It was reported that high concentrations of ascorbate were essential for adequate functioning of mononuclear and polymorphonuclear leukocytes (PMNs) and other cell types.[41-43] As pointed out earlier, both macrophages and neutrophils have been shown to accumulate ascorbic acid to high levels.[31,33] Ascorbate is depleted from these cells during infection.[44-46] In infected areas, the ascorbate accumulation by leukocytes increases with chemotoxis[47] and its concentration is diminished as a result of exposure to external agents such as steroids.[48,49] Steroids are suggested to suppress the hexose monophosphate shunt[48,40] leading to the depletion of ascorbate within the PMNs.

Many researchers have reported that the chemotactive response of phagocytes in persons with chronic disease is improved by higher intake of vitamin C. For instance, Anderson[50] showed that an intake of 1 gram of vitamin C per day in children with chronic granulomatous disease increased neutrophil mobility. Patrone and Dallegri[51] found that vitamin C, when used in patients with recurrent infections, was effective in correcting defective phagocytic function. In Chediak-Higashi patients with impaired immune systems, higher intake of vitamin C protected them against infections.[52-54]

Neutrophils are the most chemotactically responsive and they are the first to reach the inflammatory site followed by other phagocytic white blood cells. Anderson[50] reported that ascorbate injection caused an increase in neutrophil motility and lymphocyte responsiveness to mitogens. Ascorbic acid may enhance the inactivation of microorganisms in the lysosomal compartment of these cells.[55,56] In the cytoplasm, ascorbate may protect the integrity of neutrophils during oxidative burst.[57,58] Washko et al.[33] have shown that intracellular ascorbate is found in a reduced form in the cytosol and is not protein bound. In the cytosol, the vitamin may also function as an antioxidant to keep other antioxidants in a reduced form.[9,33]

Effects on T-Lymphocyte Functions

In addition to phagocytic leukocytes, ascorbate is also known to build a concentration gradient across the cell membrane of mononuclear lymphocytes.[30,59,60] Ascorbate saturation by lymphocytes has been linked to enhanced immunocompetence.[11] Several studies carried out in humans, animals and cell culture have demonstrated immunostimulatory effect of ascorbate on lymphocytic cells.[61-69] In studies with healthy human volunteers carried out by Oh and Nakano,[61] Yonemoto et al.[62] and Yonemoto,[63] it was shown that ingestion of from one to several grams a day of ascorbic acid evoked an enhancement in lymphocyte blastogenesis following exposure to pokeweed mitogens or lectins. The degree of lymphocyte blastogenesis was correlated to the amount of vitamin C administered.[64-66] Siegel and Morton reported a similar increase in T lymphocyte blastogenesis to concavalin-A following ascorbate supplementation *in vitro*.[65] Addition of ascorbic acid to cultured human lymphocytes *in vitro* also potentiated mitogen or lectin induced mononuclear DNA synthesis.[66-68] Enhanced T lymphocyte response to viral infections was also seen in healthy individuals following ingestion of several grams per day of ascorbate.[69] In addition, ascorbate ingestion was shown by Anderson[70] to inhibit the induction of suppressor activity in human volunteers.

In contrast to studies on healthy human subjects, there are relatively fewer reports on the effect of vitamin C supplementation on T cell responses in immuno-suppressed persons. Most significant among these is a clinical report from South Africa dealing with measles in native children who experience a high death rate from secondary infections.[71] These children also exhibit depressed T cell counts and abnormal T cell ratios similar to those seen in AIDS patients. Administration of a few grams per day of vitamin C during the convalescent phase was shown to restore the T cell ratio to normal levels.

Children suffering from Chediak-Higashi Disease, a rare congenital disorder characterized by drastically deranged phagocytic function, were shown to have their immune systems restored to normalcy by supplements of vitamin C.[72] A recent study from Canada showed that supplementation with a multivitamin and trace element regimen improved immune responses and reduced the rate of infections in elderly subjects.[73] Although the multivitamin-mineral supplement corrected serum ascorbate deficiency seen at baseline, the relative contribution of vitamin C in immune stimulation was difficult to assess, as only 80 mg of vitamin C were used in the Canadian study. Much larger doses of ascorbate need to be evaluated in immuno-deficient subjects and HIV-infected/AIDS patients to elucidate the relative contribution of ascorbate in immuno-enhancement.

Effect on Interferon Production

Vitamin C has been shown to stimulate endogenous production of interferons, both *in vivo* and *in vitro*.[69,74-77] Interferons are proteins produced by cells in response to viral infection or antigen stimulation that exert antiviral effects on neighboring cells. Siegel[74,75] first noted that ascorbic acid was involved in the stimulation of interferon production in two separate studies. In the first study, mice infected with leukemia virus in the presence of ascorbate showed detectable increase in interferon levels over controls.[74] In a subsequent study, a dose-dependent increase in interferon production was observed in cultured mouse cells stimulated by poly (rI):poly (rC).[75] Ascorbic acid also stimulated interferon levels in human embryonic skin and lung fibroblasts following induction by Newcastle disease virus.[76,77] The same concentrations of ascorbate did not have any effect on interferon stimulation in two lymphoblastoid cell lines induced by Sendai virus.[76] The relative contribution of endogenous interferon in immuno-enhancement by ascorbate is not known and needs to be elucidated.

ANTIBACTERIAL EFFECTS

Ascorbic acid-mediated modulation of phagocyte motility and function may aid the host's immune response against microbial infections. Thus, ascorbate-deficient guinea pigs were demonstrated to have impaired bacteriocidal activity in their polymorphonuclear leukocytes.[39,78] Chatterjee et al. found that PMNs from scorbutic animals displayed reduced lysozomal enzymatic activity and decreased capacity to phagocytize bacillus subtilis *in vitro*. These activities were restored upon supplementation with physiological concentrations of ascorbate. In other studies, no significant impairment in bacteriocidal activity against *Staphylococcus aureus* was noted either with peritoneal neutrophils or macrophages derived from scorbutic guinea pigs.[79,80] The observed differences in phagocytic activity may be related to the type of bacteria used, a result consistent with studies of ascorbic acid in humans.

Ascorbic acid has also been demonstrated to have direct bacteriostatic and bactericidal effects that date back to the 1930's. In earlier studies,[1] it was shown to be effective in the inactivation of a wide range of pathogenic bacteria which included *staphylococcus aureus, B. typhosus, E. coli, B. subtilis, B. diphtheriae* and *streptococcus hemolyticus*. Boissevain and Spillane[81] first reported on the bacteriostatic effect of ascorbic acid on *M. tuberculosis* that was subsequently confirmed by Sirsi.[82] Earlier studies also demonstrated that ascorbic acid neutralized or inactivated a wide range of bacterial toxins, such as those linked to diphtheria,[83] tetanus,[84] staphylococcal infections,[85] and dysentery.[39,86]

The mechanism by which vitamin C inactivates bacteria has been ascribed to the attack by free radicals generated as a result of the reaction of ascorbate and molecular oxygen in the presence of transition metals such as copper ions as deduced from *in vitro* models.[87-89] Murata et al.[90] evaluated the bactericidal efficacy of vitamin C using 30 different strains of bacteria and reported a wide range of susceptibility to inactivation by the vitamin. Oxygen radicals were shown to cause single-strand scission in bacterial DNA.[91] The relevance of this prooxidant *in vitro* effect of ascorbate to physiological inhibition of bacteria *in vivo* needs to be determined.

Vitamin C also plays an important role in stimulating the C1q biosynthesis, and this may affect bactericidal effects in guinea pigs. Haskell and Johnston[92] showed that C1q concentrations were higher in animals fed ascorbate concentration adequate for proper growth. Depletion of ascorbate was linked to reduced antibacterial action in neutrophils from scorbutic guinea pigs and ascorbate supplementation was shown to reverse this effect.[93]

ANTIVIRAL ACTION

A striking property of ascorbic acid is its ability to inactivate viruses and inhibit viral growth in their host cells. Ascorbic acid has been known to be an antiviral agent since 1935. Jungeblut[94] was the first to report on the inactivation of polio virus by crystalline vitamin C. As a result of this, other studies followed and demonstrated the efficacy of ascorbic acid against a wide range of viruses. These include studies by Holden and Molloy[95] on herpes viruses; Kligler and Bernkopf[96] on vaccinia virus; Langenbush and Enderling[97] on the virus which causes foot-and-mouth disease; Amato[98] on rabies; and Lojkin[99] on tobacco mosaic virus. Earlier work on the effects of vitamin C on viruses dating back to the thirties has been previously reviewed by Irwin Stone.[1]

Most of the detailed work on the mechanism of virus inactivation was carried out in subsequent studies using bacteriophages. Among the RNA phages, no two viruses were found to be alike in their sensitivity to ascorbate.

Phage MS2 was completely inactivated within 60 minutes at 37°C by 1 x 10^{-5}M ascorbate, whereas phages J1 and δA required 10-fold higher concentration for complete inactivation.[100] Further, phage R17 was resistant to inactivation by ascorbate alone and Cu^{2+} alone, whereas its infectivity was reduced by 10^6-fold in 4 minutes when incubated with ascorbate and Cu^{2+} together at concentrations of 1 x 10^{-3}M and 1 x 10^{-5}M, respectively.[100] Whether ascorbate inactivates viruses through a prooxidant effect *in vivo* is not known.

Working with a retrovirus of chickens, Bissell et al.[101] found that preincubation of avian sarcoma virus (RSV) with ascorbate *in vitro* did not reduce its infectivity. However, addition of ascorbic acid to RSV-infected cultures of primary avian tendon cells and chicken embryo fibroblasts resulted in the inhibition of tumor virus replication and cell transformation.[101] The inhibition was attributed initially to two possible mechanisms: (1) increased interferon production by RSV-infected cultures which led to reduced spread of virus and (2) production of defective virus conferring resistance to superinfection by transforming virus. However, a more recent *in vitro* study indicated an indirect effect of ascorbate on RSV replication in avian tendon cells.[102] Ascorbate was shown to stabilize the differentiated cellular phenotype in these primary cells, thereby decreasing the ability of virus to replicate and transform cell cultures. In a study of rhinovirus replication in cultured WI-38 cells, Schwerdt and Schwerdt[103] had previously observed that the reduction in virus proliferation by ascorbate was not mediated by the production of interferon.

Ascorbate has also been demonstrated to suppress human retrovirus expression in immortalized and transformed lymphocytic cell lines. Blakeslee et al.[104] showed that ascorbic acid inhibited the activation of HTLV-1 virus following induction of latently infected MT-1 cells by 5-iodo-2′-deoxyuridine and *N*-methyl-*N*′-nitro-Nitrosoguanidine. It was not determined whether the inhibitory effect of ascorbate was due to direct action of the compound on the inducer chemical or whether it involved interference with HTLV expression in stimulated cells. More recent studies with HIV carried out in our laboratory have demonstrated that ascorbate is capable of inhibiting HIV replication in both chronically and acutely infected T cell lines in the absence of inducing agents, indicating that the compound can directly interfere with specific steps in retrovirus replication in differentiated lymphocytic cells (see below).

THIOL AND ASCORBATE IMBALANCE
IN HIV INFECTION

HIV-infected persons have been reported to exhibit significantly decreased concentrations of acid-soluble thiols (cysteine and glutathione, GSH) in their body fluids and peripheral blood mononuclear cells (PBMC). Eck et al. were the first to show that the levels of total and reduced GSH in blood plasma and

PBMC of both nonsymptomatic individuals and patients with ARC or AIDS were 30% lower than the corresponding levels in healthy controls.[105] These patients also had elevated levels of glutamate in their blood plasma. Subsequent *in vitro* experiments by Eck and Dröge[106] demonstrated that high extracellular concentrations of glutamate caused a significant decrease in the intracellular content of cysteine and GSH in peritoneal macrophages. Based on these and other data showing that peritoneal macrophages and peripheral blood monocytes take up cystine and release cysteine into the extracellular space, these investigators suggested that elevated glutamate may contribute to thiol deficiency in patients with HIV/AIDS.

GSH deficiency in serum of symptom-free HIV-seropositive subjects was confirmed by Buhl et al.[107] whose data and those of Eck et al.[105] indicated that this condition was not due to a wasting syndrome. Buhl et al.[107] further showed that the GSH level in the lung epithelial lining fluid of asymptomatic persons was reduced to 60% of that in healthy subjects. Since GSH is normally present at significantly higher amounts in lungs relative to blood, and the lung is a major site of opportunistic infection, GSH deficiency may be an important factor in the pathogenesis of pulmonary infection associated with AIDS. In addition, GSH deficiency in blood may contribute to the underlying immune dysfunction. In this regard, Staal et al.[108] have shown that the decreased intracellular GSH concentrations in HIV-infected persons were due to the selective loss of high GSH-containing subpopulations of CD4 and CD8 containing T cells. It has been demonstrated that intracellular GSH is critical for certain T cell functions such as IL-2 dependent proliferation of activated T cells;[109] formation of CD8 cytotoxic T cell blasts;[110,111] activity of both cytotoxic and natural killer cells;[111] and mixed lymphocyte reactions.[109] It has been suggested that the immune dysfunction seen early in HIV infection, prior to CD4 cell depletion, may be due to GSH deficiency.[105,107,112] Studies in primates inoculated with Simian immunodeficiency virus have shown that GSH deficiency is an early event, occurring immediately as a consequence of viral infection.[113]

Other factors, in addition to viral infection, that can induce GSH depletion are xenobiotic drugs. In experimental animals that do not synthesize their own ascorbate (like humans), drug-induced GSH deficiency was shown to cause depletion of ascorbate levels in tissues followed by multiple organ damage and high mortality rate.[4] Administration of large doses of ascorbate was shown to confer cell protection and protect the animals from almost certain death due to multiple organ failure.[4,8]

GSH deficiency can also result from changes in dietary ascorbate. Volunteers fed controlled diets containing ascorbate at levels lower than the recommended daily allowance (60 mg) had decreased concentrations of GSH in blood plasma.[114] Administration of ascorbate repletion diet (250 mg) in the same study resulted in restoration of plasma GSH level.

Serum ascorbate status in HIV infection has been addressed in one study. Bogden et al. reported that 27% of HIV-infected persons examined had below normal plasma ascorbate concentrations.[26]

OXIDATIVE STRESS, INFLAMMATION, AND HIV ACTIVATION

Since HIV-infected persons are immunosuppressed, they are highly susceptible to inflammatory responses induced by oxidative stress resulting from exposure to viral/opportunistic pathogens, drugs, and other xenobiotics. Inflammation is associated with the formation of reactive oxygen intermediates (ROIs) and cytokines such as interleukins (IL-1, IL-2, IL-6), granulocyte/macrophage colony stimulative factor (GMCSF), and tumor necrosis factor alpha or beta (TNF-α or β). Evidence for oxidative damage in HIV-infected patients is supported by the detection of malondialdehyde, a by-product of lipid peroxidation.[115] HIV-infected persons have also been demonstrated to have increased levels of TNF-α and IL-6 in their blood serum.[116-118] ROIs are primarily generated during the respiratory burst of phagocytes (macrophages and neutrophils).

ROI production can also be induced by inflammatory cytokines in their target cells.[119] Increased ROI production by granulocytes has been observed during early stages of HIV infection.[120] Inflammatory cytokines also stimulate latent HIV expression in lymphocytic and monocytic cells.[121-123] This effect is mediated through the activation of the transcription factor NF-$\kappa\beta$ or via a post-transcriptional effect. TNF-α is a strong inducer of HIV, acting at the transcriptional level,[122,124] whereas IL-6 induces HIV weakly by acting at the post-transcriptional level.[125] ROIs such as hydrogen peroxide have also been shown to activate NF-$\kappa\beta$ and augment the expression of NF-$\kappa\beta$-dependent genes, such as IL-2 and TNF-α, as well as the HIV genome.[112] Because ROIs and cytokines can mutually cross-augment their respective expression and both in turn can activate HIV gene expression, a feed-back loop is generated resulting in the self-perpetuation of HIV infection. Periodic exposure to oxidative stress may create episodes of increased inflammatory response and HIV activation contributing to progressive CD4+ T cell depletion.

Cycles of increased ROI production in HIV-infected persons may subject them to consume free-radical scavengers at an increased rate, leading to depletion of vital antioxidants in the body. This mechanism is consistent with detection of lowered GSH, cysteine and ascorbate levels in blood and body fluids of HIV-infected and AIDS patients.

In well-nourished healthy individuals, antioxidants are regenerated from their oxidized forms by reducing equivalents (NADPH or FADH2) produced from oxidation of carbohydrates and fats during glycolytic/citric acid cycle.

The prooxident state in AIDS may reflect a breakdown in the mechanism to generate enough reducing equivalents to overcome the increased ROI load. Future studies aimed at analyzing sugar and fat metabolism in HIV-infected persons may throw more light on the molecular basis of antioxidant deficiency.

ASCORBATE AND THIOL EFFECTS ON HIV REPLICATION *IN VITRO*

The influence of extracellular ascorbate and thiol concentrations on HIV expression has been investigated using infected cells in culture. In chronically infected T lymphocytic cells that constitutively produce HIV, noncytotoxic ascorbate concentrations were shown to suppress HIV activity in a dose-dependant fashion. At the highest nontoxic dose, ascorbate reduced extracellular reverse transcriptase (RT) activity by 99% and p24 antigen level by over 90%.[126]

In acutely infected CD4+ T cells, nontoxic ascorbate caused a dose-dependent suppression of syncytium formation, a cytopathic marker of HIV replication. Two distinct inhibitory events were discerned in chronically infected cells: an effect on virus replication detected at early time and destabilization of virion-associated RT observed upon prolonged incubation with ascorbate. Long-term experiments showed that continuous exposure of ascorbate was necessary to keep HIV in a suppressed state.[127] More recent experiments, directed at elucidating the mechanism of action, have demonstrated that ascorbate suppresses HIV by exerting a post-translational effect targeted at inactivation of viral enzymatic activity.[128] Thus, analysis of ascorbate effect in co-cultivation assays of chronically infected and uninfected CD4+ cultures showed that the vitamin did not affect gp120 *env*-CD4 receptor interactions and did not inhibit the activity of the transactivator *tat* protein. Additionally, comparison of HIV RNA and protein patterns of ascorbate-treated cells with corresponding patterns in untreated cells did not reveal significant differences in the size or level of viral mRNAs and polypeptides, indicating an effect beyond the translational step. Analysis of the enzymatic activity of a reporter protein, expressed under the influence of HIV LTR, was reduced to ~11% of control in ascorbate-treated cells, indicating impairment of enzyme activity.[137]

Thiols such as GSH and GSH precursors (N-acetyl cysteine, NAC), have also been demonstrated to suppress HIV expression in chronically infected cells.[123,129,130] However, unlike ascorbate, the predominant inhibitory effect of thiols on extracellular HIV RT activity was seen following cytokine stimulation of latent HIV activation rather than on basal or constitutive expression.[129] When both ascorbate and NAC were tested together in a constitutively

HIV-producing cell line, a synergistic suppression of HIV RT activity was observed.[127] The mechanism underlying thiol action has been attributed to inhibition of the activation of NF-κβ, a key cellular transcription factor that mediates HIV expression directed by the viral long terminal repeat.[112,131] The effect of ascorbate on cytokine-stimulated HIV expression in latently infected cells has not been elucidated and is under investigation in our laboratory.

THERAPEUTIC STRATEGIES

Given its multifunctional properties, ascorbate offers promising possibilities as a therapeutic agent in controlling HIV infection and AIDS. The advantage ascorbate has over other anti-HIV agents is that it can be taken at large doses without producing adverse side effects. The unique ascorbate functions that become manifest at high doses, permitting it to act as a powerful scavenger of free radicals/oxidants [through direct donation of high-energy electrons], empowers it with the dual ability to keep oxidant-induced HIV activation in check as well as protect tissues from cellular damage. In addition, its ability to suppress HIV replication and inhibit microbial growth (directly or via stimulation of phagocytic cells) affords an additional line of defense against proliferation of the AIDS virus and opportunistic pathogens associated with HIV infection. Promising therapeutic strategies specifically relevant to HIV/AIDS merit discussion.

The ability of high-dose ascorbate to serve as an essential antioxidant under conditions of GSH deficiency sets it apart from all other known antioxidants. A global GSH deficiency in HIV-infected individuals indicates a state of severe free-radical/oxidant toxicity and a breakdown in the metabolic pathway that provides reducing equivalents to regenerate endogenous antioxidants from their oxidized forms (e.g., GSH disulphide $\rightarrow$ reduced GSH). Exogenous antioxidants are therefore indicated to overcome this problem. Given the high bowel tolerance for ascorbate in AIDS patients,[132] large amounts of antioxidant are required. Administration of synthetic GSH is not a viable alternative as the compound is not readily taken by cells and has proved to be toxic when administered to GSH-deficient guinea pigs.[8] Whereas GSH monoester is readily transported into cells, and has been found to be quite effective,[8] only limited quantities of it can be used as compared to ascorbate. GSH precursors, such as derivatives or prodrugs of cysteine, can provide the limiting sulfated amino acid which is used in the synthesis of GSH. However, it is unlikely that the body can tolerate the large amount of cysteine derivative that is needed to provide sufficient reducing equivalents to counteract the heavy load of free radicals and other oxidant products.

On the other hand, ascorbate can be used at high levels without adverse side effects. In this role, high-dose ascorbate has a unique function.[4,8,9] It can serve as a direct source of electrons (i.e., reducing equivalents) necessary to neutralize oxidants. As long as an adequate supply of ascorbate is maintained, there is no longer a need to convert oxidized ascorbate (DHA) back to ascorbate in GSH-deficient subjects. This unique role of ascorbate in sparing the requirement for GSH has been demonstrated in newborn rats exhibiting drug-induced GSH deficiency.[4] Large doses of ascorbate were shown to protect such animals from debilitating effects of oxygen toxicity and almost certain death due to multiple organ failure.

The mechanism of action of ascorbate on HIV also sets it apart from known anti-HIV agents, including nucleoside analogues,[133,134] alpha interferon,[134] GSH, GSH monoester and NAC.[129-131] All anti-HIV agents approved to date for AIDS treatment are reverse transcriptase inhibitors that block HIV DNA synthesis in newly infected cells but do not affect virus production in chronically infected cell lines containing integrated proviral DNA. The ability of ascorbate to suppress HIV production from chronically infected cells and interfere with viral replication in acute infection offers a distinct advantage in controlling the progression of HIV infection. Ascorbate concentrations that suppress HIV *in vitro* are attainable in blood plasma through ingestion of large doses or via intravenous infusion.[127] Other antiviral compounds that affect different steps in HIV replication may be used in combination with ascorbate. Thus, *N*-acetyl cysteine, which inhibits efficiently cytokine-stimulated HIV expression, primarily at the transcriptional level, cooperates with ascorbate in synergistic suppression of HIV production in chronically infected T cells.[127] We and others have suggested the use of ascorbate in combination with NAC for controlling progression of HIV infection.[127,131,135] In addition, other antioxidants and nutritional factors that stimulate cell-mediated immunity can be considered for combinational therapy with ascorbate. A promising candidate is beta-carotene that has been shown to elevate CD4+ counts in HIV-infected persons.[136] Clinical trials of ascorbate have been planned, including a factorial study aimed at assessing the efficacy of single and combined regimens of ascorbate and beta-carotene in HIV-infected persons and AIDS patients. Based on its mode of action, ascorbate treatment may help in slowing down the progression of HIV infection and the rate of opportunistic infection/malignancy associated with AIDS.

ACKNOWLEDGMENT

We are grateful to Professor Linus Pauling for stimulating our interest in this field. We thank Dan McWeeney for assistance with *in vitro* experiments and Martha Best for processing this manuscript.

REFERENCES

1. Stone, I., *The Healing Factor: Vitamin C against Disease,* Grosset and Dunlap, New York, NY, 258pp., 1972.
2. Lewin, S., *Vitamin C: Its Biology and Medical Potential,* Academic Press, 231pp, 1976.
3. Wells, W. W., Xu, D. P., Yang, Y. F., and Rocque, P. A., Mammalian thioltransferase (glutaredoxin) and protein disulfide isomerase have dehydroascorbate reductease activity, *J. Biol. Chem.,* 265, 15361, 1990.
4. Martensson, J., and Meister, A., Glutathione deficiency decreases tissue ascorbate levels in newborn rats: Ascorbate spares glutathione and protects, *Proc. Natl. Acad. Sci. USA,* 88, 4656, 1991.
5. Lind, J., *Treatise on the scurvy,* 3rd edition, Classics of Medicine Library, Birmingham, AL, 1771, 1980.
6. Booth, J. B., and Todd, G. B., Subclinical scurvy-hypovitaminosis C, *Geriatrics,* 27, 130, 1972.
7. Wilson, C. W. M., and Nolan, C., The diets of elderly people in Dublin, Ir., *J. Med. Sci.,* 3, 345, 1970.
8. Mårtensson, J., Han, J., Griffith, O. W., and Meister, A., Glutathione ester delays the onset of scurvy in ascorbate-deficient guinea pigs, *Proc. Natl. Acad. Sci.,* 90, 317, 1993.
9. Cathcart, R. F., A unique function for ascorbate, *Med. Hypotheses,* 35, 32, 1991.
10. Englard, S., and Seifter, S., The biochemical functions of ascorbic acid, *Annu. Rev. Nutr.,* 6, 365, 1986.
11. Cameron, E., Pauling, L., and Leibovitz, B., Ascorbic acid and cancer: A review, *Cancer Res.,* 39, 663, 1979.
12. Robertson, W. vB., and Schwartz, B., Ascorbic acid and the formation of collagen, *J. Biol. Chem.,* 201, 689, 1953.
13. Stetten, M. R., Some aspects of the metabolism of hydroxyproline, studied with the aid of isotopic nitrogen, *J. Biol. Chem.,* 181, 31, 1949.
14. Gould, B. S., and Woessner, J. F., Biosynthesis of collagen, *J. Biol. Chem.,* 226, 289, 1957.
15. Cardinale, G. J., Stassen, F. L. H., Kuttan, R., and Udenfriend, S., Activation of prolyl hydroxylase in fibroblasts by ascorbic acid, *Ann. N. Y. Acad. Sci.,* 258, 1975.
16. Stassen, F. L. H., Cardinale, G. J., and Udenfriend, S., Activation of prolyl hydroxylase in L-929 fibroblasts by ascorbic acid, *Proc. Natl. Acad. Sci. USA,* 70, 1090, 1973.
17. Levene, C. I., and Bates, C. J., Ascorbic acid and collagen synthesis in cultured fibroblasts, *Ann. N. Y. Acad. Sci.,* 258, 288, 1975.
18. McGee, J. O'D., and Udenfriend, S., Partial purification and characterization of peptidyl proline hydroxylase precursor from mouse fibroblasts, *Arch. Biochem. Biophys.,* 152, 216, 1972.
19. Peterkofsky, B., Ascorbate requirement for hydroxylation and secretion of procollagen: relationship to inhibition of collagen synthesis in scurvy, *Am. J. Clin. Nutr.,* 54, 11355, 1991.

20. Houglum, K. P., Brenner, D. A., and Chojkier, M., Ascorbic acid stimulation of collagen biosynthesis independent of hydroxylation, *Am. J. Clin. Nutr.,* 54, 1141, 1991.

21. Hughes, R. E., Hurley, R. J., and Jones, E., Dietary ascorbic acid and muscle carnitine (β-OH-γ-(trimethylamino)butyric acid) in guinea pigs, *Br. J. Nutr.,* 43, 385, 1980.

22. Hughes, R. E., Recommended daily amounts and biochemical roles — the vitamin C, carnitine, fatigue relationship, In *Vitamin C: Ascorbic Acid,* Counsell, J. N., and Hornig, D. H., eds., Applied Science Publishers, London, 75, 1981.

23. Cameron, E., Vitamin C and cancer cachexia: the carnitine connection, In *The Roots of Molecular Medicine,* Huemer, R. P., ed., W. H. Freeman & Company, New York, 133, 1986.

24. Cameron, E., Vitamin C, carnitine and cancer, *A Year in Nutritional Medicine,* Bland, J. F., ed., Keats Publishing, New Canaan, Conn., 15, 1986.

25. Cameron, E., and Pauling, L., *Cancer and Vitamin C,* Linus Pauling Institute of Science and Medicine, Menlo Park, Calif., 1979.

26. Bogden, J. D., Baker, H., Frank, O., Perez, G., Kemp, F., Breuning, K., and Louria, D., Micronutrient status and human immunodeficiency virus (HIV) infection, *Ann. N. Y. Acad. Sci.,* 587, 189, 1990.

27. Frei, B., England, L., and Ames, B. N., Ascorbate is an outstanding antioxidant in human blood plasma, *Proc. Natl. Acad. Sci. USA,* 86, 6377, 1989.

28. Prinz, W., Bortz, R., Bregin, B., and Hersch, M., The effect of ascorbic acid supplementation on some parameters of the human immunological defence system, *Int. J. Vitam. Nutr. Res.,* 47, 248, 1977.

29. Vallance, S., Relationships between ascorbic acid and serum proteins of the immune system, *Br. Med. J.,* 2, 437, 1977.

30. Evans, R. M., Currie, L., and Campbell, A., The distribution of ascorbic acid between various cellular components of blood, in normal individuals, and its relation to the plasma concentration, *Br. J. Nutr.,* 47, 473, 1982.

31. Bergsten, P., Amitai, G., Kehrl, J., Dhariwal, K. R., Klein, H., and Levine, M., Millimolar concentrations of vitamin C in purified mononuclear leukocytes: depletion and reaccumulation, *J. Biol. Chem.,* 265, 2584, 1990.

32. Washko, P., Rotrosen, D., and Levine, M., Ascorbic acid transport and accumulation in human neutrophils, *J. Biol. Chem.,* 264, 18996, 1989.

33. Washko, P. W., Rotrosen, D., and Levine, M., Transport and accumulation of ascorbic acid into human neutrophils, *Am. J. Clin. Nutr.,* 54, 1221S, 1991.

34. Feigen, G. A., Smith, B. H., Dix, C. E., Flynn, C. J., Peterson, N. S., Rosenberg, L. T., and Pavlovic, S., Enhancement of antibody production and protection against systemic anaphylaxis by large doses of vitamin C, *Res. Commun. Chem. Pathol. Pharmacol.,* 38, 313, 1982.

35. Mueller, P. S., and Kies, M. W., Suppression of tuberculin reaction in the scorbutic guinea pig, *Nature,* 195, 813, 1962.

36. Zweiman, B., Schoenwetter, W. F., and Hildreth, E. A., The effect of the scorbutic state on tuberculin hypersensitivity in the guinea pig. I. Passive transfer of tuberculin hypersensitivity, *J. Immunol.,* 96, 296, 1966.

37. Zweiman, B., Besdine, R. W., and Hildreth, E. A., The effect of the scorbutic state on tuberculin hypersensitivity in the guinea pig. II. In vitro mitotic response of lymphocytes, *J. Immunol.,* 96, 672, 1966.

38. Kalden, J. R., and Buthy, E. A., Prolonged skin allograph survival in vitamin C deficient guinea pigs, *Eur. Surg. Res.,* 4, 114, 1972.
39. Cottingham, E., and Mills, C. A., Influence of environmental temperature and vitamin deficiency on phagocytic functions, *J. Immunol.,* 477, 493, 1943.
40. DeChatelet, L. R., Cooper, R. M., and McCall, C. E., Stimulation of the hexose monophosphate shunt in human neutrophils by ascorbic acid: mechanism of action, *Antimicrob. Agents Chemother.,* 1, 12, 1972.
41. DeChatelet, L. R., McCall, C. E., Cooper, M. R., and Shirley, P. S., Ascorbic acid levels in phagocytic cells, *Proc. Soc. Exp. Biol. Med.,* 145, 1170 1974.
42. Glick, D., and Hosoda, S., Histochemistry. LXXVIII. Ascorbic acid in normal mast cells and macrophages and in neoplastic mast cells, *Proc. Soc. Exp. Biol. Med.,* 119, 52, 1965.
43. Goldschmidt, M. C., Masin, W. I., Brown, L. R., and Wyde, P. R., The effects of ascorbic acid deficiency on leukocyte phagocytosis and killing of actinomyces viscosus, *Int. J. Vitam. Nutr. Res.,* 58, 326, 1988.
44. Greene, M., and Wilson, C. W. M., Effect of aspirin on ascorbic acid metabolism during colds, *Br. J. Clin. Pharmacol.,* 2, 369, 1975.
45. Hume, R., and Weyers, E., Changes in leukocyte ascorbic acid during the common cold, *Scott. Med. J.,* 18, 3, 1973.
46. Wilson, C. W. M., and Loh, H. S., Vitamin C and colds, *Lancet,* 1, 1058, 1973.
47. Tonutti, E., and Matzner, K. H., Vitamin C bei der fermentativen Lösung toten Materials im Organismus, *Klin. Wochensch,* 17, 63, 1938.
48. Chretien, J. H., and Garagusi, V. F., Correction of corticosteroid-induced defects of polymorphonuclear neutrophil function by ascorbic acid, *J. Reticuloendoth, Soc.,* 14, 280, 1973.
49. Rivers, J. M., and Devine, M. M., Relationships of ascorbic acid to pregnancy, and oral contraceptive steroids, *Ann. N. Y. Acad. Sci.,* 258, 465, 1974.
50. Anderson, R., Assessment of oral ascorbate in three children with chronic granulomatous disease and defective neutrophil motility over a 2-year period, *Clin. Exp. Immunol.,* 43, 180, 1981.
51. Patrone, F., and Dallegri, F., Vitamin C and the phagocytic system, *Acta Vitam et Enzymol.,* 1, 5, 1979.
52. Boxer, L. A., Watanabe, A. M., Rister, M., Besch, H. R., Jr., Allen, J., and Bachner, R. L., Correction of leukocyte function in Chediak-Higashi Syndrome by ascorbate, *N. Engl. J. Med.,* 295, 1041, 1976.
53. Boxer, L. A., Vanderbilt, B., Bonsib, S., Jersild, R., Yang, H. H., and Bachner, R. L., Enhancement of chemotactic response and microtubule assembly in human leukocytes by ascorbic acid, *J. Cell. Physiol.,* 100, 119, 1979.
54. Gallin, J. I., Elin, R. J., Hubert, R. T., Fauci, A. S., Kaliner, M. A., and Wolff, S. M., Efficacy of ascorbic acid in Chediak-Higashi Syndrome: studies in humans and mice, *Blood,* 53, 226, 1979.
55. Drath, D. B., and Karnofsky, M. L., Bactericidal activity of metal-mediated peroxide-ascorbate systems, *Infect. Immun.,* 10, 1077, 1974.
56. Miller, T. E., Killing and lysis of gram-negative bacteria through the synergistic effect of hydrogen peroxide, ascorbic acid, and lysozyme, *J. Bacteriol.,* 98, 949, 1969.
57. Stankova, L., Gerhardt, N. B., Nagel, L., and Bigley, R. H., Ascorbate and phagocyte function, *Infect. Immun.,* 12, 252, 1975.

58. Anderson, R., Theron, A. J., and Ras, G. J., Regulation by the antioxidants ascorbate, cysteine, and dapsone of the increased extracellular and intracellular generation of reactive oxidants by activated phagocytes from cigarette smokers, *Am. Rev. Resp. Dis.,* 135, 1027, 1987.

59. Moser, U., Uptake of ascorbic acid by leukocytes, *Ann. N. Y. Acad. Sci.,* 498, 200, 1987.

60. Varma, S. D., Ascorbic acid and the eye with special reference to the lens, *Ann. N. Y. Acad. Sci.,* 498, 280, 1987.

61. Oh, C., and Nakano, K., Reversal by ascorbic acid of suppression by endogenous histamine of rat lymphocyte blastogenesis, *J. Nutr.,* 118, 639, 1989.

62. Yonemoto, R. H., Chretien, P. B., and Fehniger, T. F., Enhanced lymphocyte blastogenesis by oral ascorbic acid, *Proc. Am. Assoc. Cancer Res.,* 17, 288, 1976.

63. Yonemoto, R. H., In *Vitamin C: Recent Advances and Aspects of Virus Diseases and in Lipid Metabolism,* Hanck, A., ed., Huber, Vienna, 143, 1979.

64. Anderson, R., Oosthuizen, R., Maritz, R., Theron, A., and Van Rensburg, H. J., The effects of increasing weekly doses of ascorbic acid on certain cellular and humoral immune functions in normal volunteers, *Am. J. Clin. Nutr.,* 33, 71, 1980b.

65. Siegel, B. V., and Morton, J. J., Vitamin C and the immune response, *Experientia,* 33, 393, 1977.

66. Delafuente, J.C., and Panush, R. S., Modulation of certain immunologic responses by vitamin C. II. Enhancement of concanavalin A-stimulated lymphocyte responses, *Int. J. Vit. Nutr. Res.,* 50, 44, 1979.

67. Manzella, J. P., and Roberts, N. J., Human macrophage and lymphocyte responses to mitogen stimulation after exposure to influenza virus, ascorbic acid, and hyperthermia, *J. Immunol.,* 123, 1940, 1979.

68. Panush, R. S., and Delafuente, J.C. Modulation of certain immunologic responses by vitamin C, *Int. J. Vitam. Nutr. Res.,* 19, 179, 1979.

69. Panush, R. S., Delafuente, J. C., Katz, P., and Johnson, J., Modulation of certain immunologic responses by vitamin C, In *Vitamin C,* Hanck, A., ed., Wein, Huber, 35, 1982.

70. Anderson, R., Hay, I., Van Wyk, H. A., and Theron, A., Ascorbic acid in bronchial asthma, *S. Afr. Med. J.,* 63, 649, 1983.

71. Joffe, M. I., Sukha, N. R., and Rabson, A. R., Lymphocyte subsets in measles. Depressed helper inducer subpopulation reversed by in vitro treatment with levamisole and ascorbic acid, *J. Clin. Invest.,* 72, 971, 1983.

72. Baehner, R. L., Neutrophil dysfunction associated with states of chronic and recurrent infections, *Pediat. Clin. North Am.,* 27, 377, 1980.

73. Chandra, R. K., Effect of vitamin and trace-element supplementation on immune responses and infection in elderly subjects, *Lancet,* 340, 1124, 1992.

74. Siegel, B. V., Enhanced interferon response to murine leukemia virus by ascorbic acid, *Infect. Immun.,* 10, 409, 1974.

75. Siegel, B. V., Enhancement of interferon production by Poly (rl) Poly (rC) in mouse cultures by ascorbic acid, *Nature,* 254, 531, 1975.

76. Dahl, H., and Degre, M., The effect of ascorbic acid on production of human interferon and the antiviral activity in vitro, *Acta. Path. Microbiol. Immunol. Scand.,* 84, 280, 1976.

77. Karpinska, T., Kawecki, Z., and Kandefer-Szerszen, M., The influence of ultra-violet irradiation, L-ascorbic acid and calcium chloride on the induction of interferon in human embryo fibroblasts, *Arch. Immunol. Ther. Exp.,* 30, 33, 1982.

78. Chatterjee, G. C., Majumder, P. K., Banerjee, S. K., Roy, R. K., and Rudrapal, D., Relationships of proteins and mineral intake to L-ascorbic acid metabolism, including considerations of some directly related hormones, *Ann. N. Y. Acad. Sci.,* 258, 382, 1975.

79. Ganguly, R., Durieux, M. F., and Waldman, R. H., Macrophage function in vitamin C-deficient guinea pigs, *Am. J. Clin. Nutr.,* 29, 762, 1976.

80. Stankova, L., Gerhardt, N. B., Nagel, L., and Bigley, R. H., Ascorbate and phagocyte function, *Infect. Immun.,* 12, 252, 1975.

81. Boissevian, C. H., and Spillane, J. H., Effect of ascorbic acid on growth of tuberculosis bacillus, *Am. Rev. of Tuberculosis,* 35, 661, 1937.

82. Sirsi, M., Antimicrobial action of vitamin C on M. tuberculosis and some other pathogenic organisms, *Indian J. Med. Sci.,* 6, 252, 1952.

83. Sigal, A., and King, C. G., The influence of vitamin C deficiency upon the resistance of guinea pigs to diphtheria toxin, *J. Pharmacol. and Exp. Ther.,* 61, 1, 1937.

84. Jungeblut, C. W., Inactivation of tetanus toxin by crystalline vitamin C (ascorbic acid), *J. Immunol.,* 33, 203, 1937.

85. Kodama, T., and Kojima, T., Studies of the staphylococcal toxin, toxoid and antitoxin: Effect of ascorbic acid on staphylococcal lysins and organisms, *Kitasato Archv. of Exp. Med.,* 16, 36, 1939.

86. Takahashi, Z., Ein experimentelle studie über die beziehung zwischen dysenterie und vitamin C. I. Mittl: Über den einflusz des vitamin C auf meerschweinchen, denen dysenteriebazilen enigespritzt worden waren, *Nagoya J. Med. Sci.,* 12, 50, 1938.

87. DeChatelet, L. R., et al., Ascorbic Acid: Possible role in phagocytosis, Paper read at 62nd meeting, American Society of Biological Chemists, San Francisco, June 18, 1971.

88. Cooper, M. R., et al., Stimulation of leukocyte hexose monophosphate shunt activity by ascorbic acid, *Infection and Immunity,* 3, 851, 1971.

89. Erricsson, Y., and Lundbeck, H., Antimicrobial effect in vitro of the ascorbic acid oxidation. II. Influence of various chemicals and physical factors, *Acta Pathol. Microbiol. Scand.,* 37, 507, 1955.

90. Murata, A., et al., Killing effect of ascorbic acid on bacteria and yeasts, *Vitamins,* 64(12), 709, 1990.

91. Murata, A., et al., Mechanism of bactericidal action of ascorbic acid on *Escherichia coli, Vitamins,* 65(9), 439, 1991.

92. Haskell, B.E., and Johnston, C. S., Complement component C1q activity and ascorbic acid nutriture in guinea pigs, *Am. J. Clin. Nutr.,* 54, 1228S, 1991.

93. Goldschmidt, M., Reduced bactericidal activity in neutrophils from scorbutic animals and the effect of ascorbic acid on these target bacteria *in vivo* and *in vitro, Am. J. Clin. Nutr.,* 54, 1214S, 1991.

94. Jungeblut, C. W., Inactivation of poliomyelitis virus by crystalline vitamin C (ascorbic acid), *J. Exp. Med.,* 62, 517, 1935.

95. Holden, M., and Molloy, E., Further experiments on inactivation of herpes virus by vitamin C (L-ascorbic acid), *J. Exp. Med.,* 33, 251, 1937.

96. Kligler, I. J., and Bernkopf, H., Inactivation of vaccinia virus by ascorbic acid and glutathione, *Nature,* 139, 965, 1937.

97. Langenbusch, W., and Enderling, A., Einfluss der vitamine auf das virus der maul-und klavenseuch, *Zentralblatt fur Bakteriologie,* 140, 112, 1937.

98. Amato, G., Azione dell'acido ascorbico sul virus fisso della rabbia e sulla tossina tetanica, *Giornale de Batteriologia, Virologia et Immunologia,* 19, 843, 1937.

99. Lojkin, M., Contributions of the Boyce Thompson Institute, *Proceedings Third International Congress of Microbiology,* Vol. 8, No. 4, L. F. Martin, New York, 281, 1940.

100. Murata, A., and Uike, M., Mechanism of inactivation of bacteriophage M52 containing single-stranded RNA by ascorbic acid, *J. Nutr. Sci. Vitaminsol,* 22, 347, 1976.

101. Bissell, M. J., Hatie, C., Farson, D. A., Schwarz, R. I., and Soo, W., Ascorbic acid inhibits replication and infectivity of avian RNA tumor virus, *Proc. Natl. Acad. Sci. USA,* 77, 2711, 1980.

102. Schwarz, R. I., Ascorbate stabilizes the differentiated state and reduces the ability of Rous sarcoma virus to replicate and to uniformly transform cell cultures, *Am. J. Clin. Nutr.* 54, 1247S, 1991.

103. Schwerdt, P. R., and Schwerdt, C. E., Effect of ascorbic acid on rhinovirus replication in WI-38 cells (38724), *Proc. Soc. Exp. Biol. Med.,* 148, 1237, 1975.

104. Blakeslee, J. R., Yamamoto, N., and Hinuma, Y., Human T-cell leukemia virus 1 induction by 5-iodo-2'-deoxyuridine and N-methyl-N'-nitro-N-nitrosoguanidine: inhibition by retinoids, l-ascorbic acid and Dl-alpha tocopherol, *Cancer Res.,* 45, 3471, 1985.

105. Eck, H. P., Gmünder, H., Hartmann, M., Petzoldt, D., Daniel, V., and Dröge, W., Low concentrations of acid-soluble thiol (cysteine) in the blood plasma of HIV-1 infected patients., *Biol. Chem. Hoppe Seyler,* 370, 101, 1989.

106. Eck, H. P., and Dröge, W., Influence of the extracellular glutamate concentration on the intracellular cyst(e)ine concentration in macrophages and on the capacity to release cysteine, *Biol. Chem. Hoppe Seyler,* 370, 109, 1989.

107. Buhl, R., Jaffe, H. A., Holroyd, K. J., Wells, F. B., Mastrangeli, A., Saltini, C., Cantin, A. M., and Crystal, R. G., Systematic glutathione deficiency in symptom-free HIV-seropositive individuals, *Lancet,* ii, 1294, 1989.

108. Staal, F. J. T., Roederer, M., Israelski, D. M., et al., Intracellular glutathione levels in T cell subsets decrease in HIV infected individuals, *AIDS Res. Hum. Retroviruses,* 8, 305, 1992.

109. Meister, A., Glutathione metabolism and its selective modification, *J. Biol. Chem.,* 263, 17205, 1988.

110. Hamilos, D. L., and Wedner, H. J., The role of glutathione in lymphocyte activation. I. Comparison of inhibitory effects of buthionine sulfoximine and 2-cyclohexane-1-one by nuclear size transformation, *J. Immunol.,* 135, 2740, 1985.

111. Dröge, W., Pottmeyer-Gerber, C., Schmidt, H., and Nick, S., Glutathione augments the activation of cytotoxic T lymphocytes *in vivo, Immunobiology,* 172, 152, 1986.

112. Dröge, W., Eck, H. P., and Mihm, S., HIV-induced cysteine deficiency and T-cell dysfunction — a rationale for treatment with N-actylcysteine, *Immunology Today,* 13(6), 211, 1992.

113. Eck, H. P., Stahl-Henning, C., Hunsmann, G., and Dröge, W., Metabolic disorder as early consequence of simian immunodeficiency virus infection in rhesus macaques, *Lancet,* 338, 346, 1991.

114. Henning, S. M., Zhang, J. Z., McKee, R. W., Swendseid, M. E., and Jacob, R. A., Gluthathione blood levels and other oxidant defense indices in men fed diets low in vitamin C, *Am. Inst. of Nutr.,* 121, 1969, 1991.

115. Sönnerborg, A., Carlin, G., Akerlund, B., and Jarstrand, C., Increased production of malondialdehyde in patients with HIV infection, *Scand. J. Infect. Dis.,* 20, 287, 1988.

116. Breen, E. C., Rezai, A. R., Nakajima, K., Beall, G. N., Mitsuyasu, R. T., Hirano, T., Kishimoto, T., and Martinez-Maza, O., Infection with HIV is associated with elevated IL-6 levels and production, *J. Immunol.,* 144, 480, 1990.

117. Lahdevirta, J., Maury, C. P., Teppo, A. M., and Repo, H., Elevated levels of circulating cachetin/tumor necrosis factor in patients with acquired immunodeficiency syndrome, *Am. J. Med.,* 85, 289, 1988.

118. Reddy, M. M., Sorrell, S. J., Lanage, M., and Greico, M. H., Tumor necrosis factor and HIV p-24 antigen levels in serum of HIV-infected populations, *J. AIDS,* 1, 436, 1988.

119. Yoshie, O., Majima, T., and Saito, H., Membrane oxidative metabolism of human eosinophilic cell line EoL-1 in response to phorbol diester and formyl peptide: synergistic augmentation by interferon-gamma and tumor necrosis factor, *J. Leukoc. Biol.,* 45, 10, 1989.

120. Sönnerborg, A., and Jarstrand, C., Nitroblue tetrazolium (NBT) reduction by neutrophilic granulocytes in patients with HTLV-III infection, *Scand. J. Infect., Dis.,* 18, 101, 1986.

121. Folks, T. M., Justement, J., Kinter, A., Schnittman, S. M., Orenstein, J., Poli, G., and Fauci, A. S., Characterization of a promonocyte clone chronically infected with HIV and inducible by 13-phorbol-12-myristate acetate, *J. Immunol.,* 140, 1117, 1988.

122. Griffin, G. E., Leung, K., Folks, T. M., Kunkel, S., and Nabel, G. J., Activation of HIV gene expression during monocyte differentiation by induction of NF-k B, *Nature,* 339, 70, 1989.

123. Roederer, M., Staal, F. J., Raju, P. A., Ela, S. W., Herzenberg, L. A., and Herzenberg, L. A., Cytokine-stimulated human immunodeficiency virus replication is inhibited by N-acetyl-L-cysteine, *Proc. Natl. Acad. Sci. USA,* 87, 4884, 1990.

124. Nabel, G., and Baltimore, D., An inducible transcription factor activates expression of human immunodeficiency virus in T cells, *Nature,* 326, 711, 1987.

125. Poli, G., Bressler, P., Kinter, A., Duh, E., Timmer, W. C., Rabson, A., Justment, J. S., Stanley, S., and Fauci, A. S., Interleuken 6 induces human immunodeficiency virus expression in infected monocytic cells alone and in synergy with tumor necrosis factor alpha by transcriptional and post-transcriptional mechanisms, *J. Exp. Med.,* 172, 151, 1990.

126. Harakeh, S., Jariwalla, R. J., and Pauling, L., Suppression of human immunodeficiency virus replication by ascorbate in chronically and acutely infected cells, *Proc. Natl. Acad. Sci. USA,* 87, 7245, 1990.

127. Harakeh, S., and Jariwalla, R.J., Comparative study of the anti-HIV activities of ascorbate and thiol-containing reducing agents in chronically HIV-infected cells, *Am. J. Clin. Nutr.,* 54, 1231S, 1991.

128. Jariwalla, R. J., and Harakeh, S., HIV suppression by ascorbate and its enhancement by a glutathione precursor, Eighth *Int. Conf. on AIDS,* July 1992, Amsterdam, 2 B207.

129. Kalebic, T., Kinter, A., Guide, P., Anderson, M. E., and Fauci, A. S., Suppression of human immunodeficiency virus expression in chronically infected monocytic cells by glutathione ester, and N-acetyl cysteine, *Proc. Natl. Acad. Sci. USA,* 88, 986, 1991.

130. Mihm, S., Ennen, J., Pessara, U., Kurth, R., and Droge, W., Inhibition of HIV-1 replication and NF-Kb by cysteine and cysteine derivatives, *AIDS,* 497, 503, 1991.

131. Staal, F. J. T., Ela, S. W., Roederer, M., Anderson, M. T., Herzenberg, L. A., and Herzenberg, L. A., Glutathione deficiency and human immunodeficiency virus infection, *Lancet,* 339, 909, 1992.

132. Cathcart, R. F., Vitamin C in the treatment of acquired immune deficiency syndrome (AIDS), *Med. Hypotheses,* 14, 423, 1984

133. Mitsuya, H., Weinhold, K. J., Furman, P. A., St. Clair, M. H., Nusinoff-Lehrman, S., Gallo, R. C., Bolognesi, D., Barry, D. W., and Broder, S., 3'-Azido-3'-deoxythymidine (BW A509U): an antiviral agent that inhibits the infectivity and cytopathic effect of human T-lymphotropic virus type III/lymphadenopathy-associated virus *in vitro, Proc. Natl. Acad. Sci. USA,* 82, 7096, 1985.

134. Poli, G., Orenstein, J. M., Kinter, A., Folks, T. M., and Fauci, A. S., Interferon-alpha but not AZT suppresses HIV expression in chronically infected cell lines, *Science,* 244, 575, 1989.

135. Halliwell, B., and Cross, C. E., Reactive oxygen species, antioxidants and acquired immunodeficiency syndrome — sense or speculation?, *Arch. Intern. Med.,* 151, 29, 1991.

136. Coodley, G. O., Nelson, H. D., Loveless, M. O., and Folk, C., beta-carotene in HIV infection, *J. AIDS,* 6, 272, 1993.

137. Harakeh, S. and Jariwalla, R. J., Mechanistic aspects of ascorbate inhibition of human immunodeficiency virus, *Chemico-Biological Interactions,* 1994, in press.

Chapter

10

Vitamin A and HIV Infection

Brian J. Ward

and

Richard D. Semba

INTRODUCTION

Malnutrition and wasting are prominent manifestations of the late stages of HIV infection[1,2] and a growing number of single nutrient deficiencies have been reported in AIDS patients.[3-10] These changes in nutriture are unlikely to occur precipitously and there is increasing evidence that deterioration of nutritional status begins long before wasting becomes clinically apparent.[11-13] Several micronutrients are known to have profound influence on the human immune system.[14,15] Therefore, it is possible that micronutrient deficiency states can act as cofactors in HIV progression and disease expression.[16] Vitamin A has the potential to be among the most important of these cofactors.[17] Indeed, vitamin A was recognized to have potent 'anti-infective' actions[18] very shortly after its discovery as 'lipid soluble A' in 1915.[19] HIV+ individuals may be at particular risk for vitamin A deficiency for several reasons. These include chronic and recurrent infections, chronic inflammatory states, poor intake and diarrhea with or without malabsorption.[1] Vitamin A has remarkably pleotropic effects including established roles in hematopoiesis, the maintenance of epithelial integrity and optimal function of the immune system.[20-25] A central characteristic of HIV infection is the slow erosion of adaptive immunity with ever greater reliance on passive barriers and non-specific immune responses to microbial and neoplastic challenges.[1] Loss of the actions of vitamin A on epithelial tissues, non-specific immune surveillance and adaptive immunity may therefore be devastating in HIV infection, initiating a vicious spiral of

0-8493-7842-7/94/$0.00+$.50
© 1994 by CRC Press, Inc.

infection/inflammation, further depletion of vitamin A stores and ever greater susceptibility to infection, inflammation and malignancy. Although it is somewhat artificial to consider vitamin A removed from the context of the total nutriture, isolated nutritional deficiencies are common in early HIV infection and the potential for benefit from micronutrient 'intervention' is intuitively greatest before HIV infection has progressed to full-blown AIDS. In this chapter, we will briefly summarize the biology of vitamin A and then discuss the potential for interaction between changes in vitamin A status and HIV infection.

ABSORPTION, TRANSPORT, AND STORAGE OF RETINOIDS

Vitamin A is a fat-soluble substance which can be ingested either preformed (retinol) or synthesized within the body from ingested plant carotenoids.[15,26] Dietary sources of retinol are limited to animal products (fish, meat and dairy) while carotenoids are present in dark green, leafy vegetables and many brightly colored fruits and vegetables. Collectively these substances are referred to as 'retinoids'.

Retinol is esterified in the intestinal mucosa, packaged in chylomicra and delivered to the liver via the systemic circulation in chylomicron remnants.[27,28] More than 90% of the vitamin A in the body is stored in the liver as retinyl esters. Retinol is released from the liver in combination with retinol binding protein (RBP) and circulates in a trimolecular complex with transthyretin. Intestinal absorption of retinol and mobilization of retinol stores are tightly regulated to maintain constant serum retinol levels. Retinol enters target cells via surface receptors which recognize RBP and undergoes conversion in the cytosol to retinoic acid or other metabolites.[29] These metabolites are transported within the cell and to the nucleus by cellular binding proteins which vary in their tissue distribution and binding affinities.[30] At the nuclear membrane, retinoic acid (or other metabolites) associates with a growing family of specific receptors: α, β, γ retinoic acid receptors (RAR) and the newly described retinoid X receptor.[31,32] These retinoid-receptor complexes form hetero- and homodimers which act by binding directly to regulatory regions of target genes to modify gene transcription. It is in this way that the remarkable range of retinoid-induced effects is realized in different tissues and at different stages of development of the organism or the cell.[33] The uptake, handling and postulated molecular mechanisms of retinol are schematically outlined in Figures 1 and 2.

Carotenoids are divided into those which can be converted to retinol and those which cannot.[34] The best studied pro-vitamin A carotenoid is beta-carotene. Depending upon vitamin A status and the simultaneous

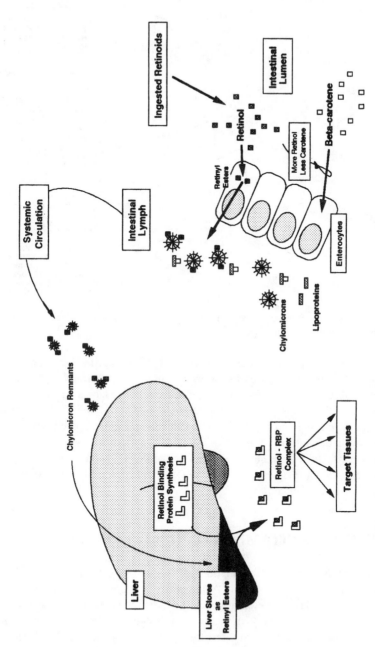

Figure 1. Schematic outline of intestinal absorption, handling, and storage of retinol.

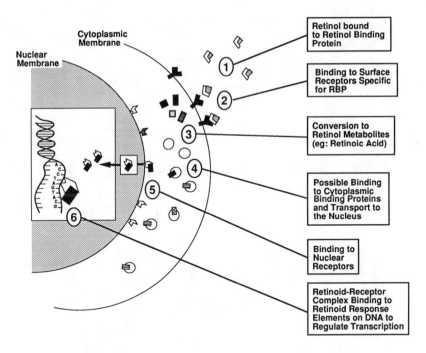

Figure 2. Schematic outline of the postulated binding, intracellular transport and molecular mechanism of action of retinol.

presence of pre-formed retinol, between 10–25% of ingested beta-carotene is converted to retinol and retinyl esters in enterocytes. Variable amounts of carotenoids are also absorbed unchanged and circulate associated with lipoproteins. Since the storage capacity for carotenoids is more limited than that for vitamin A, plasma and tissue carotenoid levels are more immediately influenced by dietary intake. Carotenoids share many actions with vitamin A but not all of the effects of these substances can be attributed to conversion to vitamin A.[35,36] See recent reviews for a more detailed discussion of the handling and molecular actions of retinoids.[28-31]

EFFECTS OF VITAMIN A ON HEMATOPOIESIS

Retinoids promote differentiation and maturation of myeloid and erythroid precursors in vitro.[26,37,38] Vitamin A deficiency is strongly associated with iron deficiency anemia in developing-world children, and supplementation results in improvement of virtually all measures of iron metabolism including hematocrit, hemoglobin, serum iron, ferritin and transferrin saturation.[39-41] Hemosiderosis of the liver has been found at autopsy in children who die with vitamin A deficiency suggesting that inability to mobilize hepatic iron stores underlies the associated anemia.[39-42] Hemoglobin levels also increase in pregnant women in response to supplemental vitamin A.[43] Disruption of normal iron metabolism has also been reported in HIV+ individuals[1,44] and anemia is an early and consistent finding in HIV infection[1,45] despite normal marrow stores in most subjects.[1] Severe anemia requiring transfusion support is not rare in advanced AIDS and hemosiderosis has occasionally been reported in HIV infection.[46] Factors thought to contribute to this anemia include direct infection of erythroid precursors in the bone marrow,[47] chronic infections and decreased production of erythropoietin.[48] The possible role of vitamin A deficiency in HIV-associated anemia is currently under investigation.

VITAMIN A DEFICIENCY IMPAIRS EPITHELIAL INTEGRITY

Epithelial tissues throughout the body are sensitive to the effects of vitamin A. In animals deprived of vitamin A, there is disruption of mucosal surfaces with loss of ciliae and abnormal microvilli in the respiratory and gastrointestinal tracts as well as the vagina and salivary glands.[42,49-55] Squamous metaplasia with loss of epithelial goblet cell mucous production is also characteristic of vitamin A deficiency in humans and has been documented in the vagina, bladder, nose, trachea, and eye.[42,54,56] The extent to which disruption of epithelial barriers contributes to the burden of microbial challenges in HIV infection is currently unknown. The effects of vitamin A on the vaginal mucosa raise the intriguing possibility that changes in vitamin A nutriture may influence heterosexual and perinatal transmission of HIV. The latter possibility is currently under investigation (J. Humphrey, personal communication).

VITAMIN A DEFICIENCY IMPAIRS IMMUNITY

The evidence that vitamin A deficiency impairs normal immune function comes from several sources including animal studies of experimental depletion and repletion,[57-72] in vitro studies of human immune cell function with supplemental retinoids,[73-82] field studies of deficiency and supplementation in children[83-93] and epidemiological studies of diet and cancer risk.[94-104]

Animals fed a diet deficient in vitamin A have multiple abnormalities in both non-specific and adaptive immunity. Impaired phagocytosis,[57] NK cell function[58] and decreased ability to recognize and eliminate tumors have all been reported.[59] Antibody production to protein antigens is poor[60-66] and is likely due to incompetent T cell help for antigen-specific IgG production.[65]

Total T helper cell numbers are decreased[65] and the capacity of thymocytes, lymphocytes and splenocytes to proliferate in response to soluble antigens and mitogens is reduced in vitamin A-deficient animals.[21,24,61,66-68] Many of these abnormalities can be improved or normalized with vitamin A supplements. The increased susceptibility of vitamin A deficient animals to a wide range of microorganisms and recovery of resistance with vitamin A supplements provides further indirect evidence of impaired immunity with deficiency of this micronutrient.[22-24,69-72]

Although the immunologic impact of changes in vitamin A status has been less thoroughly studied in man, retinoids can influence a wide range of human immune cell functions in vitro. These include proliferation of thymocytes,[73] B cells[74] and T cells,[75-78] NK activity[78] and the production of a number of cytokines.[79-82] Vitamin A-deficient children respond poorly to tetanus toxoid,[83] have low CD4+/CD8+ ratios[84] and suffer increased infection-related mortality.[85-91] Vitamin A repletion is associated with increased CD4+ counts and CD4+/CD8+ ratios[84,92] as well as improved antibody production to tetanus toxoid[84] and natural measles.[93] Large field trials of vitamin A supplementation in regions of the world where vitamin A deficiency is endemic have demonstrated that intermittent dosing can reduce childhood mortality as much as 20–50%.[85-91] This effect is presumed to be due to a reduction in infection-related deaths. The potential role for vitamin A in the prevention of cancer in humans is controversial[94,95] but a large body of epidemiologic data supports an association between low retinoid intake and increased cancer risk.[95-99] The postulated anti-cancer effects of retinoids may not be limited to enhanced immunologic detection and elimination of tumors, however. As anti-oxidants, retinoids have the potential to influence cancer induction by limiting DNA damage[100] and as anti-proliferative/differentiating agents, retinoids may slow or prevent clinical tumor expression despite the presence of transformed cells.[101] In this latter capacity, retinoids are clinically useful in the treatment of oral leukoplakia,[100,102] a pre-neoplastic condition, and promyelocytic leukemia.[103,104] The potential role for retinoids in the treatment of HIV-associated neoplastic processes (eg: Kaposi sarcoma, B-cell lymphoma[1]) is currently unknown.

MEASUREMENT OF VITAMIN A STATUS AND DEFINITION OF DEFICIENCY

The evaluation of vitamin A status can be difficult. Serum retinol levels are widely used for this purpose and are thought to provide an accurate measure of the vitamin A status of populations.[26,105] The range of 'normal' for this determination is wide however (1.05–3.50 μmol/L in adults) and single serum retinol levels may be relatively insensitive for the detection of 'deficiency' at the level of the individual.[105] Direct measurement of retinyl esters in liver biopsy specimens can be performed but has obvious limitations as a diagnostic tool. To address this problem, a number of proxies for the measurement of vitamin A status have been developed. These include conjunctival impression cytology which is based on the observation that the conjunctival epithelium is highly sensitive to vitamin A status and the relative dose response which is thought to reflect the adequacy of liver vitamin A stores.[105] Although these tests tend to be highly correlated in subjects with marginal or deficient status, the correspondence between the tests is often less than perfect.[106,107] Even liver biopsy specimens can occasionally be misleading due to regional variation in ester content within the liver.[108]

The evaluation of vitamin A status in an individual is further complicated by the fact that our concepts of 'sufficiency' and 'deficiency' of vitamin A are based largely on the observation that children with serum retinol levels below 0.70 µmol/L are at increased risk for ophthalmologic disease (night blindness, Bitot's spots, keratomalacia).[109] A 'cut-off' for biochemical deficiency of 1.05 µmol/L has been applied in several adult studies and may be appropriate since NHANES survey data show that 1.05 µmol/L is at the 2nd percentile for black men and below the 1st percentile for white men in the USA.[110] Adults with retinol levels in this range rarely have ocular signs, however. These observations and animal data strongly suggest that the requirements for vitamin A can differ widely over time and between tissues in the same individual.[17,18,20-26,111] The measurement of vitamin A status in HIV+ individuals may be particularly difficult since both vitamin A mobilization and consumption are profoundly influenced by infections and inflammatory states.[112-115] It has recently been demonstrated, for example, that many children with measles virus infection are acutely vitamin A 'deficient' and derive significant benefit from supplementation[93,116] despite liver stores which are presumed to be normal.

VITAMIN A CAN ACT AS AN IMMUNE ADJUVANT

These problems in determining vitamin A status and defining 'deficiency' may be moot since there is increasing evidence that vitamin A can act as an immune adjuvant.[117,118] In vitamin-replete animals, supplemental vitamin A[119-125] and other retinoids[126-129] can increase antibody production to protein antigens given by oral, cutaneous and intraperitoneal routes. Furthermore, retinoids can enhance lymphoproliferative,[128,129] natural killer[130,131] and CTL responses[132] as well as allograft rejection[133-136] and contact sensitization.[137] Supplemental vitamin A also appears to antagonize the immunosuppressive effects of steroids[121,138] and tolerogenic stimuli in animal models.[139] Human data are more limited but are consistent with an adjuvant role for vitamin A. Retinoids can reverse post-operative[140] and UV-induced 'immunosuppression',[141,142] increase the proportion of circulating T cells which are CD4+[92] and increase antibody responses to keyhole limpet antigen[143] and tetanus toxoid.[83] Large field trials of vitamin A supplementation in the developing world have demonstrated that children need not be clinically vitamin A deficient to derive significant benefit (ie: reduced infection-related mortality) from supplements.[85-91] These observations suggest strongly that within the limits imposed by toxicity, both vitamin A-replete and vitamin A-deficient HIV+ subjects may benefit from supplemental vitamin A.

INCREASED RISK FOR VITAMIN A DEFICIENCY IN HIV-INFECTED SUBJECTS

Vitamin deficiency states are relatively rare in the developed world. Vitamin A deficiency is particularly rare in this setting due to the abundance of retinoid-containing foods, supplementation of food staples (eg: dairy products) and the enormous storage capacity for this micronutrient in the liver.[26] Some populations are at increased risk for vitamin deficiencies, however, including intravenous drug users, alcoholics, the homeless and individuals with chronic diseases.[144] HIV-infected individuals are disproportionately represented in these risk groups.[1] In addition, several characteristics of HIV infection are likely to increase the demand for and decrease the supply of this micronutrient. These include chronic and repeated acute infections (utilization of stores), decreased oral intake (eg: oral thrush, mental status changes), active liver disease (impaired storage) and decreased intestinal absorption.[1] For example, most immunocompetent individuals establish what amounts to an armed truce with many 'endogenous' pathogens such as the herpes viruses (eg: CMV, EBV) after the initial infection has been controlled. Declining immune surveillance may allow these organisms and other normally 'low-grade' pathogens to challenge the immune system of HIV+ individuals with ever greater success long before symptomatic AIDS develops. The 'baseline' state of immune activity is elevated in asymptomatic HIV+ subjects as measured by such markers as plasma neopterin (an indicator of monocyte activation)[145] and the soluble forms of the IL-2 receptor and CD8 molecules (indicators of T cell activation).[146] One of the metabolic 'costs' of this chronic state of immune activation is the consumption of vitamin A. In later stages of HIV infection, decreased food intake due to oral or esophageal infections, inanition or mental status changes may contribute to vitamin A deficiency.[1] Although intractable diarrhea and wasting are common complications of full-blown AIDS,[1] significant malabsorption may be present in a surprising number of asymptomatic HIV+ individuals, possibly as a result of direct infection of enterocytes by HIV[147] or clinically inapparent infection with enteric pathogens.[148]

VITAMIN A STATUS IN HIV INFECTION

The impact of HIV infection on vitamin A status has been examined in a number of studies in recent years with consistent results.[4-6,149,150] Mean serum retinol levels for HIV+ adults fall in the normal range, yet a significant minority have retinol levels below 1.05 μmol/L, the cut-off for biochemical deficiency. In a micronutrient survey of 112 asymptomatic homosexuals, Baum et al found biochemical vitamin A deficiency in 16%.[5] Although the authors stated that the proportion of individuals with 'multiple vitamin deficiencies' increased over time, no data are provided to follow vitamin A status changes with HIV progression. Their observations could not be explained by decreased vitamin A intake since none of the study subjects consumed less than the recommended daily allowance (RDA) for vitamin A[151] and >95% were consuming from 2.5 to more than 25 times the RDA for this micronutrient. Similar findings have been reported by the same group in a study of 100 homosexuals and 42 non-homosexuals with HIV infection.[6] The mean serum retinol for all subjects was again in the normal range but 18% of the HIV+ group had levels below 1.05 μmol/L. Bogden et al studied a small number of subjects at various stages in HIV infection (N = 30).[4] Once again, the mean serum retinol level was normal and 12% of the subjects were biochemically deficient in vitamin A. Carotenoid levels were also measured in this study and a significant minority (31%) had subnormal levels. Retinol and carotenoid levels were lower in AIDS patients than in subjects with asymptomatic HIV infection but the differences did not reach significance. Perhaps reflecting the limited storage capacity for carotenes compared to vitamin A, none of the subjects in this study had carotene levels higher than normal whereas 12% of HIV+ subjects had supranormal retinol levels. Although the total number of subjects in these studies is relatively small, the data suggest an association between HIV infection and vitamin A deficiency and are consistent with a deterioration of vitamin A and carotene status with advancing HIV. A single study of vitamin A status in HIV+ children has been reported in abstract form with similar results.[149] Mean serum retinol and RBP levels were normal but individual levels were not reported.

Since the evaluation of vitamin A status using a single measure can be problematic, we assessed vitamin A status in 25 HIV+ individuals with CD4+ counts ≤800 mm^{-3} using a battery of tests.[139] These included serum retinol and RBP determinations, conjunctival impression cytology (CIC) as a measure of end-organ response to available vitamin A, and relative dose responses (RDR) as a measure of vitamin A stores.[18] Of the subjects, 5 had clinically stable AIDS, 2 had diarrhea and weight loss and 18 were asymptomatic. In agreement with previous studies, the mean serum retinol (2.06 ± 0.2 μMol/L) and RBP (8 ± 1 units/mL) levels fell within the normal ranges but a significant minority (12%) had serum retinol levels below 1.05 μmol/L. When all of the tests were evaluated, 8/25 subjects (32%) had ≥1 test suggestive of vitamin A deficiency: serum retinol levels <1.05 μmol/L (3), abnormal CIC (6), abnormal RDR (2), RBP <7 (1) (Figure 3). Analyzed by stage of HIV infection, 4/5 (80%) of AIDS patients and 4/18 (22%) of asymptomatic HIV+ subjects had at least some evidence of vitamin A deficiency. Subjects with abnormal CIC tended to have lower serum retinol levels than those with normal CIC (1.54 ± 0.30 vs 2.20 ± 0.21 μmol/L; P = 0.15) and 60% of subjects with AIDS had abnormal CIC (vs 17% in non-AIDS subjects; Fisher's exact test; P <0.04). Retinol levels in subjects with CD4+ counts ≤300 tended to be lower than in those with CD4+ counts >300 (1.80 ± 0.24 vs 2.20 ± 0.24 μmol/L; P = 0.15). It is also interesting that the subjects in this study who took a daily multivitamin containing only modest amounts of vitamin A (5–7000 IU) had significantly higher serum retinol levels than those taking no supplements (2.50 ± 0.24 vs 1.47 ± 0.21 μmol/L; P <0.009). Although the number of subjects in this study was also small and the correspondence between the various measures of vitamin status was limited, these data add to the evidence that vitamin A status is abnormal in a significant proportion of HIV+ subjects and suggest strongly that vitamin A status deteriorates with advancing HIV infection. The normal mean serum retinol levels that we and other investigators have observed likely reflect a shift in the distribution of retinol values in HIV+ populations. Some HIV-infected individuals take large doses of vitamin supplements (skewing to high values) and others have deteriorating vitamin A status (skewing to lower values) while the mean value for the population as a whole remains 'normal'. Bogden and co-workers noted supranormal serum retinol levels in 12% of their subjects[4] and 2/18 (11%) of the asymptomatic subjects in our study had serum retinol levels above the normal range.[150]

POSSIBLE CONSEQUENCES OF VITAMIN A DEFICIENCY IN HIV INFECTION

The preceding discussion and data demonstrate that the potential for vitamin A deficiency to act as a cofactor in HIV disease progression or expression is real. There is limited but increasing evidence that this is the case. Two small studies have reported effects of beta-carotene supplementation in HIV+ subjects which may cautiously be interpreted as beneficial. Garewal et al gave a small daily dose of beta-carotene (60 mg/day) to 11 HIV+ subjects for 4 months.[152] Although there was no change in total T cell number or CD4+/CD8+ ratios, the number of cells expressing NK markers was increased. Coodley et al have recently reported the results of substantially more aggressive supplementation.[153] In their study, 21 HIV+ subjects received 180,000 mg of beta-carotene/day for 4 weeks in a placebo-controlled, double-blind, crossover trial. Beta-carotene treatment was associated with increased total

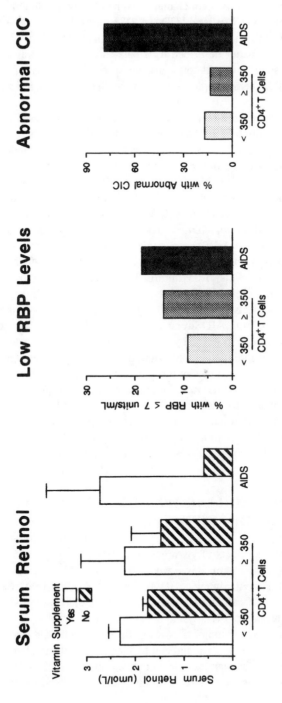

Figure 3. Deterioration of vitamin A status in HIV+ subjects: 18 asymptomatic and 5 clinically stable patients with AIDS. Seven asymptomatic and three AIDS subjects were taking a daily multivitamin.

leukocyte counts and percent changes (vs the placebo treatment period) in the proportion of CD4+ T cells and CD4+/CD8+ ratios. Despite the small numbers of subjects in this study and the fact that one half of the cross-over 'placebo' group might be better described as a beta-carotene 'withdrawal' group, the findings are suggestive that beta-carotene treatment can have a positive impact on several of the hematologic parameters which normally deteriorate with HIV progression.

The consequences of vitamin A deficiency superimposed on HIV infection are potentially severe. If a malignant interaction exists between vitamin A deficiency and HIV infection, the effect of such an interaction should be apparent in increased morbidity, decreased survival or both. We have recently completed a cross-sectional study of vitamin A status with longitudinal follow-up for mortality in a stratified sample of 179 subjects from the more than 2000 intravenous drug users in the ALIVE study (AIDS Linked to Intravenous Experiences, Baltimore, MD).[154] The study population consisted of 53 HIV-seronegative and 126 HIV-seropositive subjects. Two of the 53 seronegative individuals (3.7%) and 19 of the 126 seropositive individuals (15.1%) had plasma vitamin A levels <1.05 μmol/L. There was a strong association between low serum retinol levels and absolute CD4 counts for both seropositive and seronegative subjects. The mean follow-up was 22.8 months during which time 15 subjects died from AIDS-related causes. Kaplan-Meier product limit estimates for subgroups stratified by vitamin A status (above or below 1.05 μmol/L) revealed that individuals with vitamin A deficiency had significantly decreased mean survival time by the log-rank test (P <0.0001)(Figure 4). The adjusted odds ratio for serum retinol as an independent predictor of mortality was 4.3 compared to 10.0 for CD4+ cell counts <200 mm^{-3} and 10.1 for HBsAg$^+$. Increased infant mortality has also been observed in children born to HIV+ mothers with vitamin A deficiency in Rwanda.[155] These findings suggest strongly that vitamin A status is an important cofactor in HIV progression.

POSSIBLE MECHANISM(S) OF RETINOID-INDUCED BENEFIT IN HIV-INFECTED SUBJECTS

Many of the effects of vitamin A on epithelial tissues and immune cells outlined in the discussion above would be considered to be beneficial in HIV infection. Of particular interest, HIV infection is characterized by a progressive decline in CD4+ T cell numbers[1] and decreased capacity to produce the T cell cytokines interleukin-2 (IL-2) and interferon-γ (IFN-γ).[156,157] We have demonstrated that the proportion of CD4+ T cells and CD4+/CD8+ ratios are low in vitamin A deficient children and can be increased with supplementation.[84] As noted above, similar changes in leukocyte numbers and PBMC phenotypes are observed following beta-carotene supplementation in HIV+ subjects.[153] We

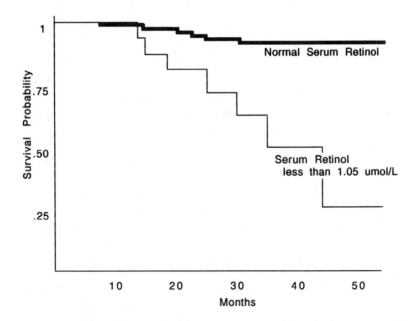

Figure 4. Kaplan Meier survival curves for 126 HIV+ individuals stratified according to serum retinol status with longitudinal follow-up for a mean of 22.8 months. (Published with permission from Archives of Internal Medicine, 1993; 153: 2148–2154.)

have also recently demonstrated that production of IL-2 and IFN-γ by PBMC isolated from HIV+ subjects can be increased with brief exposure to physiologic concentrations of retinoids *in vitro*.[150] Retinoids may contribute to the transcriptional control of IL-2 and IFN-γ[158,159] and these cytokines are essential for cellular immune responses to intracellular parasites to which HIV-infected individuals are particularly susceptible.[1,160,161] Production of IL-2 and IFN-γ is impaired early in HIV infection[156,157] and it has recently been suggested that the capacity to produce these cytokines may be a critical factor in HIV progression.[162] We found that increased cytokine production following retinoid exposure *in vitro* was independent of serum retinol levels in our HIV+ subjects suggesting that lymphocytes isolated from biochemically vitamin A-'replete' subjects can still be retinoid responsive. Finally, it has recently been suggested that HIV-induced apoptosis of immune cells may contribute to the immunologic abnormalities characteristic of this illness.[163] Apoptosis is an active process of programmed cell death which has been demonstrated in thymocytes, lymphocytes and monocytes/macrophages.[164] Possible mechanisms of HIV-induced apoptosis include direct infection of immune cells, actions of the envelope or other viral proteins and/or generalized immune activation.[163] Retinoids have been demonstrated to inhibit activation-induced apoptosis of murine T cell hybridomas, thymocytes and monocyte-like cells.[165,166] The impact of retinoids on HIV-induced apoptosis of immune cells is currently under investigation.

CAUTIONS FOR RETINOID SUPPLEMENTATION IN HIV INFECTION

The potential for retinoid-induced toxicity has recently been the subject of several comprehensive reviews.[167-171] At very high doses, vitamin A may be teratogenic and both acute and chronic toxicity have been reported. Acute hypervitaminosis A in adults can follow ingestion of foods containing high concentrations of pre-formed vitamin A (eg: liver from carnivores or large fish) or large doses of vitamin A supplements. Manifestations of acute vitamin A toxicity include abdominal pain, desquamation, visual disturbances and increased intracranial pressure with nausea, headache and papilledema. The individual susceptibility to chronic vitamin A toxicity varies widely, and cases have been reported in adults following daily ingestion of as little as 25,000 IU for prolonged periods of time (many months to years).[168] Chronic hypervitaminosis A can result in anorexia, weight loss, bone pain and liver damage. Protein deficiency, alcohol consumption, ingestion of vitamin D and co-existing liver or end-stage renal disease may all increase the risk of toxic effects from the regular ingestion of even modest amounts of vitamin A.[168,172] Since the conversion of carotenoids to vitamin A is under tight control, even pro-vitamin A carotenoids can be ingested in large amounts without significant risk of hypervitaminosis A. Except for yellow-orange discoloration of the skin, hypercarotenemia has not been associated with significant toxicity.

It is theoretically possible that supplemental retinoids may benefit HIV itself or HIV-related pathogens. The potential for such a competition between host and microbe for a limiting micronutrient is best illustrated by the enhanced virulence of a wide spectrum of human pathogens if the supply of available iron is increased.[173-176] Indeed, the *in vitro* data support this concern for retinoid supplementation in HIV infection. Several groups have reported increased HIV replication[177-181] or HIV long terminal repeat transcription[182,183] in human monocytes and monocyte-like cells in the presence of supplemental retinoid. The *in vivo* data are limited but argue against a deleterious effect of retinoid supplementation in HIV+ individuals. Survival of MAIDS-infected mice is improved with retinoic acid supplementation[184] and retinoids have been used to treat severe skin diseases in HIV-infected subjects without obvious acceleration of HIV progression.[185-187] In a study of 288 HIV+ homosexual men in Baltimore, dietary and supplemental intake of vitamin A (9,000–20,000 IU/day) was associated with slower progression to AIDS during a seven-year follow-up (A. Tang, personal communication). Finally, our data demonstrating an association between low serum retinol levels and risk of mortality in HIV+ subjects[154] suggest strongly that if HIV benefits from improved vitamin A status, the simultaneous benefit to the host is even greater.

CONCLUSION

Our understanding of vitamin A is rapidly expanding; its measurement, mechanism(s) of action, influence on a wide range of tissues and impact on infectious, inflammatory and neoplastic conditions are all areas of active research. Despite the current limitations in the measurement of vitamin A status, it is important to consider the possibility of vitamin A deficiency in HIV+ subjects since a significant minority (15–22%) of even asymptomatic subjects will be biochemically 'deficient'. Studies to identify risk factors for deficiency are currently underway and treatment of these individuals with vitamin A supplements is appropriate (eg: 200,000 every 2 months). The potential role for supplemental retinoids as adjuvant therapy in vitamin A-replete HIV+ individuals is currently supported primarily by animal data and compelling logic. The available human data are consistent with a retinoid-induced benefit, however, and clinical trials are in progress to explore this possibility more fully.

The potential role for retinoids as adjuvant therapy in HIV is perhaps most exciting in HIV-infected populations of the developing world. In these settings, the risk of vitamin deficiencies of all kinds is great and the range of available treatments for HIV is limited. The cost of a bimonthly 200,000 IU capsule of vitamin A (approximately $0.12) compares favorably with all other interventions in HIV infection. While unlikely to be a panacea, the demonstration of a beneficial effect of vitamin A supplementation in HIV infection may offer the hope of treatment where there is currently none.

REFERENCES

1. Chaisson, D.E. and Volberding, P.A., Clinical manifestations of HIV infection. In *Principles and Practise of Infectious Diseases*. Mandell, G.L., Douglas, R.G. Jr., and Bennett, J.E., (eds): Churchill Livingston, New York, pp 1059, 1990.
2. Hecker, A.M. and Kotler, D.P., Malnutrition in patients with AIDS, *Nutr. Rev.,* 48, 393, 1990.
3. Beck, K.W., Schramel, P., Hedl, A., Jaeger, H., and Kaboth, W., Serum trace element levels in HIV-infected subjects, *Biol. Trace. Element. Res.,* 25, 89, 1990.
4. Bogden, J.D., Baker, H., Frank, O., Perez, G., Kemp, F., Bruening, K., and Louria, D., Micronutrient status and human immunodeficiency virus (HIV) infection, *Ann. N.Y. Acad. Sci.,* 587, 189, 1990.
5. Baum, M.K., Shor-Posner, G., Bonveni, P., Cassetti, I., Mantero-Atienza, E., Beach, R.S., and Sauberlich, H.E., Influence of HIV infection on vitamin status and requirements, *Ann. N.Y. Acad. Sci.,* 669, 165, 1992.

6. Beach, R.S., Mantero-Atienza, E., Shor-Posner, G., Javier, J.J., Szapocznik, J., Morgan, R., Sauberlich, H.E., Cornwell, P.E., Eisdorfer, C., and Baum, M.K., Specific nutrient abnormalities in asymptomatic HIV-1 infection, *AIDS*, 6, 701, 1992.

7. Smith, J., Howell, D.W., Kendall, B., Levinsky, R., and Hyland, K., Folate deficiency and demyelination in AIDS, *Lancet*, 2, 2315, 1987.

8. Burkes, R.I., Cohen, H., Krailo, M., Snow, R.M., and Carmel, R., Low serum cobalamin levels in the acquired immunodeficiency syndrome and related disorders, *Eur. J. Haematol.*, 38, 141, 1987.

9. Foresti, V. and Confanolieri, F., Wernickes encephalopathy in AIDS, *Lancet*, 1, 1499, 1987.

10. Falutz, J., Tsoukas, C., and Gold, P., Zinc as a cofactor in human immunodeficiency virus-induced immunosuppression, *JAMA*, 259, 2850, 1988.

11. Kotler, D., Wang, J., and Pierson, R.N., Body composition studies in patients with the acquired immunodeficiency syndrome, *Am. J. Clin. Nutr.*, 42, 1255, 1985.

12. Kotler, D.P., Therney, A.R., Wang, J., and Pierson, R.N. Jr., Magnitude of body-cell-mass depletion and the timing of death from wasting in AIDS, *Am. J. Clin. Nutr.*, 50, 444, 1989.

13. Ott, M., Lembcke, B., Fischer, H., Jager, R., Polat, H., Geier, H., Rech, M., Shaszeswki, S., Helm, E.B., and Caspary, W.F., Early changes in body composition in human immunodeficiency virus-infected patients: tetrapolar body impedance analysis indicates significant malnutrition, *Am. J. Clin. Nutr.*, 57, 15, 1993.

14. Beisel, W.R., Vitamins and the immune system, *Ann. N.Y. Acad. Sci.*, 587, 5, 1990.

15. Barker, B.M., Vitamin A. In *Vitamins in Medicine*. Barker, B.M. and Bender, D.A. (eds): Vol 2. Fourth edition. London: Heinemann, pp 211, 1983.

16. Beach, R.S., Mantero-Atienza, E., Van Riel, F., and Fordyce-Baum, M., Potential implications of nutritional deficiencies in early HIV-1-infected patients, *Arch. AIDS Res.*, 3, 225, 1989.

17. Scrimshaw, N.S., Synergistic and antagonistic effect of nutrition and immunity, *Fed. Proc., Fed. Am. Soc. Exp. Biol.*, 25, 1679, 1966.

18. Green, H.N. and Mellanby, E., Vitamin A as an anti-infective agent, *BMJ*, 2, 691, 1928.

19. McCollum, E.V. and Davis, M., The essential factors in the diet during growth, *J. Biol. Chem.*, 23, 231, 1915.

20. Jurin, M. and Tannock, I., Influence of vitamin A on immunological responses, *Immunology*, 45, 3435, 1980.

21. Chandra, R.K. and Au, B., Single nutrient deficiency and cell-mediated immune responses. III. Vitamin A, *Nutr. Res.*, 1, 181, 1981.

22. Watson, R.R. and Rybsky, J.A., Immunologic response modification by vitamin A and other retinoids. In *Nutrition and Immunology*, New York, Alan R. Liss, pp 87, 1988.

23. West, C.E., Rombout, J.H.W.M., Van der Zijpp, A.J., and Sijtsma, S.R., Vitamin A and immune function, *Proc. Nutr. Soc.*, 50, 251, 1991.

24. Humphery, J.H. and West, K.P. Jr., Vitamin A deficiency: Role in childhood infection and mortality. In *Micronutrients and Health*. Bendich, A. and Butterworth, C.E. Jr. (eds): New York, Marcel Dekker, pp 307, 1991.
25. Ross, C., Vitamin A status: Relationship to immunity and antibody response, *Proc. Am. Soc. Exp. Biol.,* 200, 303, 1992.
26. Olsen, J.A., Physiologic and metabolic basis of major signs of vitamin A deficiency. In *Vitamin A Deficiency and Its Control.* Baurenfeind, J.C. (ed): Orlando, Academic Press, pp 19, 1986.
27. Norum, K.R. and Blomhoff, R., McCollum Award Lecture, 1992: Vitamin A absorption, transport, cellular uptake and storage, *Am. J. Clin. Nutr.,* 56, 735, 1992.
28. Blomhoff, R., Green, M.H., Green, J.B., Berg, T., and Norum, K.R., Vitamin A metabolism: New perspective on absorption, transport and storage, *Physiol. Rev.,* 71, 951, 1992.
29. Editorial. Cell-surface receptors for retinol-binding protein, *Nutr. Rev.,* 49, 218, 1991.
30. Wolf, G., The intracellular vitamin A-binding proteins: An overview of their functions, *Nutr. Rev.,* 49, 1, 1991.
31. Carson-Jurica, M.A., Schrader, W.T., and O'Malley, B.W., Steroid receptor superfamily: structure and functions, *Endocr. Rev.,* 11, 201, 1990.
32. Lehmann, J.M., Fanjul, A., Cameron, J.F., Lu, X.P., Haefner, P., Dawson, M.I., and Pfahl, M., Retinoids selective for retinoid X receptor response pathways, *Science,* 258, 1944, 1992.
33. Levin, A.A., Sturzenbecker, L.J., Kazmer, S., Bosakowski, T., Huselton, C., Allenby, G., Speck, J., Kratzeisen, C.L., Rosenberg, M., Lovely, A., and Grippo, J.F., A new pathway for vitamin A. Understanding the pleotropic effects of retinoids, *Ann. N.Y. Acad. Sci.,* 669, 70, 1992.
34. Bendich, A. and Olson, J.A., Biological actions of carotenoids, *FASEB J.,* 3, 1927, 1989.
35. Bendich, A., Beta-carotene and the immune response, *Proc. Nutr. Soc.,* 50, 263, 1991.
36. Prabhala, R.H., Garewal, H.S., Hicks, M.J., Sampliner, R.E., and Watson, R.R., The effects of 13-cis-retinoic acid and beta-carotene on cellular immunity in humans, *Cancer,* 67, 1556, 1991.
37. Hodges, R.E., Sauberlich, H.E., Canham, J.E., Rucker, R.B., Meija, L.A., and Mohanram, M., Hematopoietic studies in vitamin A deficiency, *Am. J. Clin. Nutr.,* 31, 876, 1978.
38. Mohanran, M., Kuikarni, K.A., and Reddy, V., Hematological studies in vitamin A deficient children, *Int. J. Vitamin. Nutr. Res.,* 47, 389, 1977.
39. Mejia, L.A., Hodges, R.E., Arroyave, G., Viteri, F., and Torun, B., Vitamin A deficiency and anemia in Central American children, *Am. J. Clin. Nutr.,* 30, 1175, 1977.
40. Mejia, L.A. and Arroyave, G., The effect of vitamin A fortification of sugar on iron metabolism in preschool children in Guatemala, *Am. J. Clin. Nutr.,* 36, 87, 1982.
41. Bloem, M.W., Wedel, M., van Agtmaal, E.J., Speek, A.J., Saowakontha, S., and Scheurs, W.H.P., Vitamin A intervention: short-term effects of a single, oral, massive dose on iron metabolism, *Am. J. Clin. Nutr.,* 51, 76, 1990.

42. Blackfan, K.D. and Wobach, S.B., Vitamin A deficiency in infants: a clinical and pathological study, *J. Pediatr.*, 3, 679, 1933.

43. Panth, M., Shatrugna, V., Yasodhara, P., and Sivakumar, B., Effect of vitamin A supplementation on haemoglobin and vitamin A levels during pregnancy, *Br. J. Nutr.*, 64, 351, 1990.

44. Harris, C.E., Biggs, J.C., Concannon, A.J., and Dodds, A.J., Peripheral blood and bone marrow findings in patients with acquired immunodeficiency syndrome, *Pathology*, 22, 206, 1990.

45. Mancuso, S., Gentile, G., Perricone, R., Cajozzo, A., and Abbadessa, V., Therapeutic effect of recombinant human erythropoietin (rHuEPO) on HIV-induced anemia and zidovudine-induced anemia, *Int. Conf. AIDS*, 8, B213 (abst PoB 3734), 1992.

46. Gherardi, R.K., Mhiri, C., Baudrimont, M., Roullet, E., Berry, J.P., and Poirier, J., Iron pigment deposits, smallvessel vasculitis and erythrophagocytosis in the muscle of human immunodeficiency-infected patients, *Hum. Pathol.*, 22, 1187, 1991.

47. Folks, T.M., Kessler, S.W., Orenstein, J.M., Justement, J.S., Jaffee, E.S., and Fauci, A.S., Infection and replication of HIV-1 in purified progenitor cells of normal human bone marrow, *Science*, 242, 919, 1988.

48. Miles, S.A., Erythropoietin in the treatment of anemia in AIDS patients: Results of a multicenter, double-blind, placebo-controlled study, *Symposium Proceedings. V Int Conf AIDS*, 5, 19, 1989.

49. De Luca, L., Maestri, N., Bonanni, F., and Nelson, D., Maintenance of epithelial cell differentiation: The mode of action of vitamin A, *Cancer*, 30, 1326, 1972.

50. McDowell, E.M., Keenan, K.P., and Huang, M., Effects of vitamin A deprivation on hamster tracheal epitheliaum, *Virchows Arch. (Cell. Pathol.)*, 45, 1, 1984.

51. Rojanapo, W., Lamb, A., and Olson, J.A., The prevalence, metabolism and migration of goblet cells in rat intestine following induction of rapid, synchronous vitamin A deficiency, *J. Nutr.*, 110, 178, 1980.

52. Strum, J.M., Latham, P.S., Schmidt, M.C., and McDowell, E.M., Vitamin A deprivation in hamsters, *Virchows Arch. (Cell. Pathol.)*, 50, 43, 1985.

53. Wong, Y.C. and Buck, R.C., An electron microscopic study of metaplasia of the rat tracheal epithelium in vitamin A deficiency, *Lab. Invest.*, 24, 55, 1971.

54. Wolbach, S.B. and Howe, P.R., Tissue changes following deprivation of fat-soluble A vitamin, *J. Exp. Med.*, 42, 753, 1920.

55. Zile, M.H., Bunge, E.C., and De Luca, H.F., Effect of vitamin A deficiency on intestinal cell proliferation in the rat, *J. Nutr.*, 107, 552, 1977.

56. Keenum, D.G., Semba, R.D., Wirasasmira, S., Natadisastr, G., Muhilal, West, K.P. Jr., and Sommer, A., Assessment of vitamin A status by disk applicator for conjunctival impression cytology, *Arch. Ophthalmol.*, 108, 1436, 1990.

57. Ongaskul, M., Sirisinha, S., and Lamb, A.J., Impaired blood clearance of bacteria and phagocytic activity in vitamin A deficient rats (41999), *Proc. Soc. Exp. Biol. Med.*, 178, 204, 1985.

58. Bowman, T.A., Goonewardene, I.M., Pasatiempo, A.M.G., Ross, A.C., and Taylor, C.E., Vitamin A deficiency decreases natural killer cell activity and interferon production in rats, *J. Nutr.*, 120, 1264, 1990.

59. Seifter, E., Rettura, G., and Levenson, S.M., Decreased resistance of C3H/HeHa mice to C3HBA tumor transplants: increased resistance due to supplemental vitamin A, *J. Natl. Cancer Inst.,* 67, 467, 1981.

60. Smith, S.M., Levy, N.S., and Hayes, C.E., Impaired immunity in vitamin A-deficient mice, *J. Nutr.,* 117, 857, 1986.

61. Smith, S.M. and Hayes, C.E., Contrasting impairments in IgM and IgG responses of vitamin A deficient mice, *Proc. Natl. Acad. Sci.,* 84, 5878, 1987.

62. Pasatiempo, A.M.G., Bowman, T.A., Taylor, C.E., and Ross, A.C., Vitamin A depletion and repletion: effects on antibody response to the capsular polysaccharide of streptococcus pneumoniae, type III (SSS-III), *Am. J. Clin. Nutr.,* 49, 501, 1989.

63. Frieman, A., Sklan, D., Impaired T lymphocyte immune response in vitamin A depleted rats and chicks, *Br. J. Nutr.,* 62, 439, 1989.

64. Pasatiempo, A.M.G., Abaza, M., Taylor, C.E., and Ross, A.C., Effects of timing and dose of vitamin A on tissue retinol concentrations and antibody production in the previously vitamin A-depleted rats, *Am. J. Clin. Nutr.,* 55, 443, 1992.

65. Carman, J.A., Smith, S.M., and Hayes, C.E., Characterization of a helper T lymphocyte defect in vitamin A-deficient mice, *J. Immunol.,* 142, 388, 1989.

66. Kinoshita, M., Pasatiempo, A.M.G., Taylor, C.E., and Ross, A.C., Antibody production in vitamin A-depleted rats is impaired after immunization with bacterial polysaccharide or protein antigens, *FASEB J.,* 4, 2518, 1990.

67. Nauss, K.M., Phua, C.C., Ambrogi, L., and Newberne, P.M., Immunologic changes during progressive stages of vitamin A deficiency in the rat, *J. Nutr.,* 115, 909, 1985.

68. Butera, S. and Krakowka, S., Assessment of lymphocyte function during vitamin A deficiency, *Am. J. Vet. Res.,* 47, 850, 1986.

69. Krishnan, S., Krishnan, A.D., Mustafa, A.S., Talwar, G.P., and Ramalingaswami, V., Effect of vitamin A and undernutrition on the susceptibility of rodents to a malarial parasite, Plasmodium berghei, *J. Nutr.,* 106, 784, 1976.

70. Nauss, K.M., Anderson, C.A., Conner, M.W., and Newberne, P.M., Ocular infection with herpes simplex virus (HSV-1) in vitamin A deficient and control rats, *J. Nutr.,* 115, 1300, 1985.

71. Friedman, A., Meidovsky, A., Leitner, G., and Sklan, D., Decreased resistance and immune response to E. Coli in chicks with low and high intakes of vitamin A, *J. Nutr.,* 121, 395, 1991.

72. Carman, J.A., Pond, L., Nashold, F., Wassom, D.L., and Hayes, C.E., Immunity to *Trichinella spiralis* infection in vitamin A-deficient mice, *J. Exp. Med.,* 175, 111, 1992.

73. Sidell, N., Famatiga, E., and Golub, S.H., Augmentation of human thymocyte proliferative responses by retinoic acid, *Exp. Cell. Biol.,* 49, 239, 1981.

74. Buck, B.J., Ritter, G., Dannecker, L., Katta, V., Cohen, S.L., Chait, B.T., and Hammerling, U., Retinol is essential for growth of activated human B cells, *J. Exp. Med.,* 171, 1613, 1990.

75. Moriguchi, S., Jackson, J.C., and Watson, R.R., Effects of retinoids on human lymphocyte function in vitro, *Human Toxicol.,* 4, 365, 1985.

76. Abb, J. and Deinhardt, F., Effects of retinoic acid on the human lymphocyte response to mitogens, *Exp. Cell. Biol.,* 48, 169, 1980.

77. Dillehay, D.L., Cornay, W.J. III, Walia, A.S., and Lamon, E.W., Effects of retinoids on human thymus-dependent and thymus independent mitogenesis, *Clin. Immunol. Immunopathol.*, 50, 100, 1989.

78. Goldfarb, R.H. and Herberman, R.B., Natural killer cell reactivity: regulatory interactions among phorbol ester, interferon, cholera toxin and retinoic acid, *J. Immunol.*, 126, 2129, 1982.

79. Soppi, E., Tertti, R., Soppi, A-M., Toivanen, A., and Jansen, C.T., Differential in vitro effects of etretinate and retinoic acid on the PHA and CON A induced lymphocyte transformation, suppressor cell induction and leukocyte migration inhibitory factor (LMIF) production, *J. Immunopharmacol.*, 4, 437, 1982.

80. Matikainen, S., Serkkola, E., and Hurme, M., Retinoic acid enhances IL-1 beta expression in myeloid leukemia cells and in human monocytes, *J. Immunol.*, 147, 162, 1990.

81. Sidell, N. and Ramsdell, F., Retinoic acid upregulates interleukin-2 receptors on activated human thymocytes, *Cell. Immunol.*, 115, 299, 1988.

82. Abril, E.R., Rybski, J.A., Scuderi, P., and Watson, R.R., Beta-carotene stimulates human leukocytes to secrete a novel cytokine, *J. Leukocyte Biol.*, 45, 255, 1989.

83. Semba, R.D., Muhilal, Scott, A.L., Natadisastra, G., Wirasasmita, S., Mele, L., Ridwan, E., West, K.P. Jr., and Sommer, A., Depressed immune response to tetanus in children with vitamin A deficiency, *J. Nutr.*, 122, 101, 1992.

84. Semba, R.D., Muhilal, Ward, B.J., Griffin, D.E., Scott, A.L., Natadisastra, G., West, K.P. Jr., and Sommer, A., Altered T cell subsets in vitamin A deficiency, *Lancet*, 341, 5, 1993.

85. Daulaire, N.M.P., Starbuck, E.S., Houston, R.M., Church, M.S., Stukel, T.A., and Pandey, M.R., Childhood mortality after high dose of vitamin A in a high risk population, *BMJ*, 304, 207, 1992.

86. Kothari, G., The effect of vitamin A prophylaxis on morbidity and mortality among children in urban slums of Bombay, *J. Trop. Ped.*, 37, 141, 1991.

87. Muhilal, Permeisih, D., Idjradinata, Y.R., Muherdiyantiningsih, and Karyadi, D., Vitamin A-fortified monosodium glutamate and health, growth and survival of children: a controlled field trial, *Am. J. Clin. Nutr.*, 48, 1271, 1988.

88. Rahmathullah, L., Effect on morbidity and mortality of a modest improved intake of vitamin A, *New Engl. J. Med.*, 323, 929, 1990.

89. Sommer, A., Tarwotjo, I., Djunaedi, E., West, K.P. Jr., Loeden, A.A., Tilden, R., and Mele, L., Aceh Study Group, Impact of vitamin A supplementation on child mortality. a randomized community trial, *Lancet*, 1, 1169, 1986.

90. Fauzi, W.W., Chalmers, T.C., Herrera, M.G., Mosteller, F., Vitamin A supplementation and child mortality. A meta-analysis, *JAMA*, 269, 898, 1993.

91. West, K.P. Jr., Pokhrel, R.P., Katz, J., LeClerq, S.C., Khatry, S.K., Shrestha, S.R., Pradhan, E.K., Teilsch, J.M., Pandey, M.R., and Sommer, A., Efficacy of vitamin A in reducing preschool child mortality in Nepal, *Lancet*, 338, 67, 1991.

92. Alexander, M., Newmark, H., and Miller, G., Oral beta-carotene can increase the number of OKT4 positive cells in human blood, *Immunology*, 9, 221, 1985.

93. Coutsoudis, A., Broughton, M., and Coovadia, H.N., Vitamin A supplementation reduces measles morbidity in young African children: a randomized, placebo-controlled, double blind trial, *Am. J. Clin. Nutr.*, 54, 980, 1991.

94. Willett, W.C., Polk, B.F., Underwood, B.A., Stampfer, M.J., Pressel, S., Rosner, B., Taylor, J.O., Schneider, K., and Hames, C.G., Relation of serum vitamins A and E and carotenoids to the risk of cancer, *New Engl. J. Med.,* 310, 430, 1984.
95. Peto, R., Doll, R., Buckley, J.D., Sporn, M.B., Can dietary beta-carotene materially reduce human cancer rates?, *Nature,* 290, 201, 1981.
96. Ames, B.N., Dietary carcinogens and anticarcinogens, *Science,* 221, 1256, 1983.
97. Watson, R.R., Retinoids and vitamin E: Modulation of immune functions and cancer resistance. In *Vitamins and Cancer — Human Cancer Prevention by Vitamins and Micronutrients,* Humana Press Inc, pp 439, 1985.
98. DeCosse, J.J., Potential for chemoprevention, *Cancer,* 50, 2550, 1982.
99. Shibata, A., Paganini-Hill, A., Ross, R.K., and Henderson, B.E., Intake of vegetables, beta-carotene, vitamin C and vitamin supplements and cancer incidence among the elderly: a prospective study, *Br. J. Cancer,* 60, 673, 1992.
100. Garewal, H.S., Potential role of beta-carotene and antioxidant vitamins in the prevention of oral cancer, *Ann. N.Y. Acad. Sci.,* 669, 260, 1992
101. Hicks, R.M., The scientific basis for regarding vitamin A and its analogues as anti-carcinogenic agents, *Proc. Nutr. Soc.,* 42, 83, 1983.
102. Garewal, H.S., Meyskens, F.L., Killen, D., Reeves, D., Kiershc, T.A., Elletson, H., Strosberg, A., King, D., and Steinbronn, K., Response of oral leukoplakia to beta-carotene, *J. Clin. Oncol.,* 8, 1715, 1990.
103. Warrell, R.P., Frankel, S.R., Miller, W.H., Scheinberg, D.A., Itri, L.M., Hittelman, W.N., Vyas, R., Andreeff, M., Tafuri, A., Jakubowski, A., Gabrilove, J., Gordon, M.S., and Dmitrovsky, E., Differentiation therapy of acute promyelocytic leukemia with tretinoin (all-trans-retinoic acid), *New Engl. J. Med.,* 324, 1385, 1991.
104. Dulaney, A.M. and Murgatroyd, R.J., Use of trans-retinoic acid in the treatment of acute promyelocytic leukemia, *Ann. Pharmacother.,* 27, 211, 1993.
105. Underwood, B.A., Biochemical and histological methodologies for assessing vitamin A status in human populations, *Methods in Enzymol.,* 190, 242, 1990.
106. Underwood, B.A., Seigel, H., Weisell, R.C., and Dolenski, M., Liver stores of vitamin A in a normal population dying suddenly or rapidly from unnatural causes in New York City, *Am. J. Clin. Nutr.,* 23, 1037, 1970.
107. Solomons, N.W., Morrow, F.D., Vasquez, A., Bulux, J., Guerrero, A.M., and Russell, R.M., Test-retest reproducibility of the relative dose response for vitamin A status in Guatemalan adults: issue of diagnostic sensitivity, *J. Nutr.,* 120, 738, 1990.
108. Amédée-Manesme, O., Furr, H.C., and Olson, J.A., The correlation between liver vitamin A concentrations in micro- (needle biopsy) and macrosamples of human liver specimens obtained at autopsy, *Am. J. Clin. Nutr.,* 39, 315, 1984.
109. Sommer, A., *Nutritional Blindness: Xerophthalmia and Keratomalacia,* New York, Oxford University Press, 1982.
110. National Council for Health Statistics, Hematological and nutritional biochemistry reference data for persons 6 months — 74 years of age: United States 1976–1980. Data from NHANES I survey, NCHS Series 11, No. 232.
111. Biesalski, H.K. and Stofft, E., Biochemical, morphological and functional aspects of systemic and local vitamin A deficiency in the respiratory tract, *Ann. N.Y. Acad. Sci.,* 669, 325, 1992.

112. Scrimshaw, N.S., Taylor, C.E., and Gordon, J.E., *Interaction of Nutrition and Infection*. Geneva, World Health Organization, 1968.

113. Arroyave, G. and Calcano, M., Descenso de los niveles sericos de retinol y de su proteina de enlace (RBP) durante las infecciones, *Arch. Latinam. Nutr.*, 29, 233, 1979.

114. Campos, F.A., Flores, H., and Underwood, B.A., Effect of an infection on vitamin A status in children as measured by the RDR, *Am. J. Clin. Nutr.*, 46, 91, 1987.

115. Harris, A.D. and Moore, T., Vitamin A in infective hepatitis, *BMJ*, 1, 553, 1947.

116. Frieden, T.R., Sowell, A.L., Henning, K.J., Huff, D.L., and Gunn, R.A., Vitamin A levels and severity of measles: New York City, *Am. J. Dis. Child*, 146, 182, 1992.

117. Spitznagel, J.K. and Allison, A.C., Mode of action of adjuvants: retinol and other lysosome-labilizing agents as adjuvants, *J. Immunol.*, 104, 119, 1970.

118. Dennert, G., Immunostimulation by retinoic acid, In *The Retinoids*, Ciba Foundation Symposium, 113, 117, 1985.

119. Dresser, D.W., Adjuvancy of vitamin A, *Nature*, 217, 527, 1968.

120. Dresser, D.W., An assay for adjuvancy, *Clin. Exp. Immunol.*, 3, 877, 1968.

121. Cohen, B.E. and Cohen, I.K., Vitamin A: adjuvant and steroid antagonist in the immune response, *J. Immunol.*, 111, 1376, 1973.

122. Falchuk, K.R., Walker, W.A., Perrotto, J.L., and Iselbacher, K.J., Effect of vitamin A on the systemic and local antibody responses to intragastrically administered bovine serum albumin, *Infect. Immun.*, 17, 361, 1977.

123. Jurin, M. and Tannock, I., Influence of vitamin A on immunological response, *Immunology*, 45, 3435, 1980.

124. Barnett, J.B. and Bryant, B.L., Adjuvant properties of retinol on IgE production in mice, *Int. Arch. Allergy Appl. Immunol.*, 1979, 59, 69-74.

125. Charabati, M.F. and McLaren, D.S., Action des differentes formes actives de la vitamine A sur le mechanismes immunitaire chez le rat, *Experientia*, 29, 343, 1972.

126. Barnett, J.B., Immunomodulating effects of 13-cis retinoic acid on the IgG and IgM response of BALB/c mice, *Int. Arch. Allergy Appl. Immunol.*, 72, 227, 1983.

127. Barnett, J,B., Immunopotentiation of the IgE response by 13-cis retinoic acid, *Int. Arch. Allergy Appl. Immunol.*, 67, 287, 1982.

128. Friedman, A., Induction of immune response to protein antigens by subcutaneous co-injection with water miscible vitamin A derivatives, *Vaccine*, 9, 122, 1991.

129. Athanassiades, T.J., Adjuvant effect of vitamin A palmitate and analogs on cell-mediated immunity, *J. Natl. Cancer Inst.*, 67, 1153, 1981.

130. Dennert, G. and Lotan, R., Effects of retinoic acid on the immune system: stimulation of killer cell induction, *Eur. J. Immunol.*, 8, 23, 1978.

131. Dennert, G., Crowley, C., Kouba, J., and Lotan, R., Retinoic acid stimulation of the induction of mouse killer T-cells in allogeneic and syngeneic systems, *J. Natl. Cancer Inst.*, 62, 89, 1979.

132. Lotan, R. and Dennert, G., Stimulatory effects of vitamin A analogues on induction of cell-mediated cytotoxicity in vivo, *Canc. Res.*, 39, 55, 1979.

133. Medawar, P.B., Hunt, R., and Mertin, J., An influence of diet on transplantation immunity, *Proc. R. Soc. Lond. B*, 206, 265, 1979.

134. Medawar, P.B. and Hunt, R., Anti-cancer actions of retinoids, *Immunology*, 42, 349, 1981.
135. Malkovsky, M., Edwards, A.J., Hunt, R., Palmer, L., and Medewar, P.B., T-cell mediated enhancement of host-vs-graft reactivity in mice fed a diet enriched in vitamin A acetate, *Nature (Lond)*, 320, 338, 1983.
136. Floersheim, G.L. and Bollag, W., Accelerated rejection of skin homografts by vitamin A acid, *Transplant*, 15, 564, 1972.
137. Miller, K., Maisey, J., and Malkovsky, M., Enhancement of contact sensitization in mice fed a diet enriched in vitamin A acetate, *Int. Arch. Allergy Appl. Immunol.*, 75, 120, 1984.
138. Nuwayri-Salti, N. and Murad, T., Immunologic and anti-immunosuppressive effects of vitamin A, *Pharmacol*, 30, 181, 1985.
139. Malkovsky, M., Medawar, P.B., Hunt, R., Palmer, L., and Dore, C., A diet enriched in vitamin A acetate or the in vivo administration of interleukin-2 can counteract a tolerogenic stimulus, *Proc. R. Soc. Lond. B Biol. Sci.*, 220, 439, 1984.
140. Cohen, B.E. and Cullen, P.R., Reversal of postoperative immunosuppression in man by vitamin A, *Surg. Gyn. Obst.*, 149, 658, 1979.
141. Fuller, C., Faulkner, H., Bendich, A., Parker, R.S., and Roe, D.A., Effect of B-carotene supplementation on photosuppression of delayed type hypersensitivity in normal young men, *Am. J. Clin. Nutr.*, 56. 684, 1992.
142. Matthews-Roth, M.M., Pathak, M.A., Parrish, J., Fitzpatrick, T.B., Kass, E.H., Toda, K., and Clemens, W., A clinical trial of the effects of oral beta carotene on the responses of human skin to solar radiation, *J. Invest. Dermatol.*, 59. 349, 1972.
143. Sidell, N., Connor, M.J., Chang, B., Lowe, N.J., and Borok, M., Effects of 13-cis retinoic acid therapy on human antibody responses to defined protein antigens, *J. Invest. Dermatol.*, 95, 597, 1990.
144. Wilson, J.D., Vitamin deficiency and excess, In Harrison's Principles and Practice of Internal Medicine, Wilson, J.D. et al (eds): New York, McGraw Hill, pp 434, 1991.
145. Fuchs, D., Hausen, A., Reibnegger, G., Werner, E.R., Dierich, M.P., and Wachter, H., Neopterin as a marker for activated cell-mediated immunity, *Immunol. Today*, 9, 150, 1988.
146. Griffin, D.E., Immunologic abnormalities accompanying acute and chronic infection, *Rev. Infect. Dis.*, 13 (Suppl 1), S129, 1991.
147. Zeitz, M., Ullrich, R., Heise, W., Bergs, C., et al., Malabsorption is found in early stages of HIV infection and independent of secondary infections, *Int. Conf. AIDS*, 7, 46 (abst # WB 90), 1991.
148. Laughton, B.E., Druckman, D.A., Vernon, A., Quinn, T.C., Polk, B.F., Modlin, J.F., Yolken, R.H., and Bartlett, J.G., Prevalence of enteric pathogens in homosexual men with and without acquired immunodeficiency syndrome, *Gastroenterology*, 94, 984, 1988.
149. Calle, P.F., Miguel, M.J.G., Codoceo, R., Gomez, A., Mellado, M.J., Fontelo, P.M., and de Jose, M.I., Status of vitamin A and its carrier in pediatric population infected with HIV, *VIIth Int. Conf. AIDS*, 7, Abstract no 2508, 1991.
150. Ward, B.J., Humphrey, J.D., Clement, L., and Chaisson, R.E., Vitamin A status in HIV infection, *Nutr. Res.*, 13, 157, 1993.

151. Committee on the tenth edition of the RDA, In *Fat Soluble Vitamins*. Food and Nutrition Board, National Research Council, National Academy of Sciences, Chapter 7, pp 78, 1989.

152. Garewal, H.S., Ampel, N.M., Watson, R.R., Prabhala, R.H., and Dols, C.L., A preliminary trial of beta carotene in subjects with human immunodeficiency virus, *J. Nutr.*, 122, 728, 1992.

153. Coodley, G.O., Nelson, H.D., Loveless, M.O., and Folk, C., Beta-carotene in HIV infection, *J. AIDS*, 6, 272, 1993.

154. Semba, R.D., Graham, N.M.H., Caiaffa, W.T., Margolick, J.B., Clement, L., and Vlahov, D., Increased mortality associated with vitamin A deficiency during HIV-1 infection, *Arch. Intern. Med.*, 1993 (in press).

155. Dushimimana, A., Graham, N.M.H., Humphrey, J.H., et al., Maternal vitamin A levels in HIV-related birth outcome in Rwanda, *VIII Int. Conf. AIDS*, 8, (abstr), 1992.

156. Murray, H.M., Welte, K., Jacobs, J.L., Rubin, B.Y., Mertelsmann, R., and Roberts, R.B., Production of and in vitro response to IL-2 in AIDS, *J. Clin. Invest.*, 76, 1959, 1985.

157. Murray, H.W., Rubin, B.Y., Masur, H., and Roberts, R.B., Impaired production of lymphokines and immune (gamma) interferon in the acquired immunodeficiency syndrome, *New Engl. J. Med.*, 310, 883, 1984.

158. Felli, M.P., Vacca, A., Meco, D.A., Screpanti, I., Farina, A.R., Maroder, M., Martinotti, S., Petrangeli, E., Frati, L., and Gulino, A., Retinoic acid-induced down-regulation of the interleukin-2 promoter via cis-regulatory sequences containing an octamer motif, *Mol. Cell. Biol.*, 11, 4771, 1991.

159. Blalock, J.E. and Giford, G.E., Retinoic acid (vitamin A acid) induced transcriptional control of interferon production, *Proc. Natl. Acad. Sci. U.S.A.*, 74, 5382, 1977.

160. Smith, K., Interleukin-2. Inception, impact and implications, *Science*, 240, 1160, 1988.

161. Nathan, C.F., Prendergast, T.J., Wiebe, M.E., Stanley, E.R., Platzer, E., Remold, H.G., Welte, K., Rubin, B.Y., and Murray, H.W., Activation of human macrophages: Comparison of other cytokines with interferon gamma, *J. Exp. Med.*, 160, 600, 1984.

162. Clerici, M. and Shearer, G.M., A Th1-Th2 switch is a critical step in the etiology of HIV infection, *Immunol. Today*, 14, 107, 1993.

163. Laurent-Crawford, A.G., Krust, B., Muller, S., Riviere, Y., Rey-Cuille, M-A., Bechet, J-M., Montagnier, L., and Hovanessian, A.G., The cytopathic effect of HIV is associated with apoptosis, *Virology*, 185, 829, 1991.

164. Cohen, J.J. and Duke, R.C., Apoptosis and programmed cell death in immunity, *Annu. Rev. Immunol.*, 10, 267, 1992.

165. Iwata, M., Mukai, M., Nakal, Y., and Iseki, R., Retinoic acids inhibit activation induced apoptosis in T cell hybridomas and thymocytes, *J. Immunol.*, 149, 3302, 1992.

166. Oritani, B.K., Kaisho, T., Nakajima, K., and Hirano, T., Retinoic acid inhibits interleukin-6-induced macrophage differentiation and apoptosis in a murine hemopoietic cell line, Y6, *Blood*, 80, 2298, 1992.

167. Bauernfeind, J.C., The safe use of vitamin A: A report of the International Vitamin A Consultative Group (IVACG), Nutrition Foundation, Washington, DC, 1980.
168. Bendich, A. and Langseth, L., Safety of vitamin A, *Am. J. Clin. Nutr.,* 49, 358, 1989.
169. Biesalski, H.K., Comparative assessment of the toxicology of vitamin A and retinoids in man, *Toxicology,* 57, 117, 1989.
170. Hathcock, J.N., Hattan, D.G., Jenkins, M.Y., McDonald, J.T., Sundaresan, P.R., and Wilkening, V.L., Evaluation of vitamin A toxicity, *Am. J. Clin. Nutr.,* 52, 183, 1990.
171. Geubel, A.P., de Galoscy, C., Alves, N., Rahier, J., and Dive, C., Liver damage caused by therapeutic vitamin A administration: estimate of dose-related toxicity in 41 cases, *Gastroenterology,* 100, 1701, 1991.
172. Shmunes, E., Hypervitaminosis A in a patient with alopecia receiving renal dialysis, *Arch. Dermatol.,* 115, 882, 1979.
173. Barry, D.M.J. and Reeve, A.W., Increased incidence of Gram-negative neonatal sepsis with intramuscular iron administration, *Pediatrics,* 60, 908, 1977.
174. Murray, M.J., Murray, A.B., Murray, M.B., and Murray, C.J., The adverse effects of iron repletion on the course of certain infections, *BMJ,* 2, 1113, 1978.
175. Mazzoleni, G., De Sa, D., Gately, J., and Riddell, RH., *Yesinia enterocolitica* infection with ileal perforation associated with iron overload and deferoxamine therapy, *Dig. Dis. Sci.,* 36, 1154, 1991.
176. Abe, F., Inaba, H., Katch, T., and Hotchi, M., Effects of iron and desferroxamine on Rhizopus infection, *Mycopathologia,* 110, 87, 1990.
177. Yamamoto, N., Harada, S., and Nakashima, H., Substances affecting the infection and replication of human immunodeficiency virus (HIV), *AIDS Res.,* 2, S183, 1986.
178. Kitano, K., Baldwin, G.C., Raines, M.A., and Golde, D.W., Differentiating agents facilitate infection of myeloid leukemia cell lines by monocytotropic HIV-1 strains, *Blood,* 76, 1980, 1990.
179. Poli, G., Kinter, A.L., Justement, J.S., Bressler, P., Kehrl, J.H., and Fauci, A.S., Retinoic acid mimics transforming growth factor beta in the regulation of human immunodeficiency virus expression in monocytic cells, *Proc. Natl. Acad. Sci. U.S.A.,* 89, 2689, 1992.
180. Poli, G. and Fauci, A.S., The effects of cytokines and pharmacologic agents on chronic HIV infection, *AIDS Res. Hum. Retrovirus,* 8, 191, 1992.
181. Turpin, J.A., Vargo, M., and Meltzer, M.S., Enhanced HIV-1 replication in retinoid-treated monocytes, *J. Immunol.,* 148, 2539, 1992.
182. Collins, S.J., Retinoic acid-induced differentiation of retrovirus-infected HL-60 cells is associated with enhanced transcription from the viral long terminal repeat, *J. Virol.,* 62, 4349, 1988.
183. Maio, J. and Brown, F.L., Regulation of expression driven by human immunodeficiency virus type 1 and human T-cell lymphotropic virus type 1 long terminal repeats in pluripotential human embryonic cells, *J. Virol.,* 62, 1398, 1988.
184. Watson, R.R., Yahya, D., Darban, H.R., and Prabhala, R.H., Enhanced survival by vitamin A supplementation during a retrovirus infection causing murine AIDS, *Life Sci.,* 43, 13, 1988.

185. Martin, A.G., Weaver, C.C., Cockerell, C.J., and Berger, T.G., Pityriasis rubra pilaris in the setting of HIV infection: clinical behaviour and association with explosive cystic acne, *Br. J. Dermatol.*, 126, 617, 1992.
186. Williams, H.C. and Du Vivier, A.W.P., Etretinate and AIDS-related Reiter's disease, *Br. J. Dermatol.*, 124, 389, 1991.
187. Blauvert, A., Nahass, G.T., Pardo, R.J., and Kerdel, F.A., Pityriasis rubra pilaris and HIV infection, *J. Am. Acad. Dermatol.*, 24, 703, 1991.

Chapter

11

Effect of HIV on Metabolism and the Relationship to Muscle and Body Protein Content

T.P. Stein

INTRODUCTION

Five factors are likely to be contributors to the net loss of muscle protein in AIDS patients. Four of the factors are metabolic in origin and are the subject of this review. First, and the most important, is an energy deficit, either due to increased energy expenditure or decreased intake. Three other mechanisms are likely to be contributors to the net loss of muscle protein breakdown, (1) hypermetabolism, (2) the role of muscle as a source of glutamine to support the immune system in stressed states and (3) atrophy secondary to disuse. A fourth possibility, of non-metabolic origin, is muscle loss secondary to the effects of the AIDS virus on the nervous system. AIDS can cause neuromuscular disorders and myopathy.[16,20,21] Since there is no known nutritional or metabolic component to the neuromuscular aspects of the disease, the topic will not be discussed further.

Nearly all of the studies on protein wasting in AIDS patients have focused on the nutritional aspects, and this is probably correct; a chronic ongoing negative energy and protein balance will inevitably lead to death. Very little work has as yet been done on the potentially important areas of the relationship between muscle as a substrate source for the immune system or atrophy secondary to reduced activity. Therefore this review is concerned mainly with

0-8493-7842-7/94/$0.00+$.50
© 1994 by CRC Press, Inc.

the nutritional aspects of protein wasting, but also includes a discussion of the potential role of glutamine and inactivity induced atrophy in muscle wasting.

NUTRITIONAL STATUS OF AIDS PATIENTS

Poor nutrition and weight loss frequently occur in AIDS patients.[11,15,43,45,53,54,60] This is serious because many studies have shown that poor nutritional status is generally associated with decreased survival. Quantitatively, the major site of protein loss is skeletal muscle because it is by far the largest single protein pool in the body. Changes in lean body mass primarily effect changes in muscle protein content.

Other smaller systems also are impacted by protein deficiency. Of particular importance to the AIDS patient are the immune, respiratory and gastrointestinal systems. Undernutrition results in decreased immunocompetence[10,17,40] and deterioration of the respiratory system.[4,78] Thus *pneumocystis carinii* infection is common in both malnourished children in undeveloped countries and in AIDS patients.[25,36] The gut, which has recently been shown to play an important role in the immune system, also atrophies with undernutrition.[17,67]

BODY COMPOSITION

Part of the rationale for being concerned with nutritional status is to determine whether nutrition is a limiting factor in the body's ability to function normally and respond to the disease and any associated infections. Assessment of nutritional status encompasses the measurement of total body protein, energy reserves (principally fat), vitamins, minerals and micronutrients. The emphasis is primarily on energy stores and body protein. The most important component is the body protein status. Assessment of body fat content is not a reliable index of wasting.[31] Kotler found that in some AIDS patients loss of lean body mass (protein) could occur with little accompanying loss of body fat.[46] When AIDS patients are described as being malnourished or nutritionally depleted, the inference is that they are protein- and usually energy-depleted. Tables 1 and 2 show the magnitude of the protein depletion.

The reason for protein being so important is because of its role in the body. Protein serves both as the machinery and as part of the structure of the body. All of the metabolic functions in the body, from cell division to obtaining energy from foodstuffs to host defense mechanisms, are effected by proteins. In addition, some proteins, such as collagen and elastin, are involved in structural support of the body. The role of the other nutrients (fats, carbohydrates, vitamins and minerals) is primarily, although not exclusively, to support protein metabolism. Death inevitably results when body protein losses approach 30–40%. In AIDS patients, weight loss was an excellent predictor of

TABLE 1.
Nutritional Status of Patients 61 Consecutive Patients With AIDS

Characteristics	Mean
Age (years)	35.7 ± 1.0
Weight (kg)	63.6 ± 1.2
Ideal wt (kg)	68.8 ± 0.8
Percent ideal wt (kg)	93.9 ± 1.7
Usual wt (kg)	74.9 ± 1.6
Weight loss (%)	15.0 ± 1.4
Period of wt. loss (months)	12.5 ± 2.2
Serum Albumin (g/l)	32 ± 10
Dietary Intake	
Energy intake	1108 ± 83
Calc. needs (kcal/d)	1585 ± 25
Protein intake (g)	53 ± 4.1
Calc. needs (g)	83 ± 1.2

Adapted from Trujillo, W.B., et al., *J. Am. Diet. Assoc.,* 92, 477, 1992.

TABLE 2.
Nutritional Status of a Select Group of Asymptomatic Patients With AIDS and Apparently Normal Intakes

	AIDS	Homosexuals	Heterosexuals
Wt (kg)	57.7 ± 2.5	65.4 ± 3.3	72.8 ± 4.7*
Body cell mass (kg)	20.8 ± 0.6	25.7 ± 1.5	26.6 ± 1.5*
% ideal body wt.	97.0 ± 4 4	96.3 ± 3.0	104 ± 7.4*
Body fat (kg)	7.6 ± 1.9	9.3 ± 1.4	11.7 ± 3.8
Body K(mmol)	3126 ± 96	3855 ± 226	3989 ± 219*
Body water (L)	37.0 ± 1.1	43.0 ± 2.4	47.0 ± 2.7*
RMR (kcal)	1224 ± 126	1724 ± 54	1920 ± 72*
Energy intake (kcal/d)	2478 ± 52	2452 ± 190	2603 ± 322

* $p < 0.05$ vs heterosexuals.

Adapted from Kotler, D.P., et al., *Am. J. Clin. Nutr.,* 51, 7, 1990.

time of death, with death occurring at about 66% of ideal body weight.[44] This relationship is similar to that found with cancer patients. The close parallel suggests that it is the consequences of the protein depletion rather than the particular disease which causes death.[31]

Assessment of body protein content is difficult because of the necessity of making the measurements non-invasively with minimal inconvenience to the subject and at low cost. Of the major components of the body, total body protein is the hardest to measure. Only neutron activation analysis gives a direct measurement and can therefore give an accurate value for muscle protein mass. Total body potassium measures lean body mass. Other methods are even less direct, inferring protein status from measurements of body fat or water content.

The more rigorous methods, such as NMR, neutron activation analysis, total body potassium and total body water by isotope dilution, tend to be technically complex, expensive and hence impractical. The data are, however, reliable. At the other extreme are anthropomorphic measurements based on such measurements as body weight (especially changes in body weight) and skinfolds. These measurements give reasonable assessments of the status of body composition, particularly if applied to large groups but need to be treated with caution when applied to an individual. To obtain reliable data on an individual using these methods, it is advisable to make several measurements (e.g. anthropometrics, weight, plasma proteins etc.) and draw conclusions from the overall picture. In between are controversial methods such as TOBEC and bioelectric impedance. Although they have been extensively validated in healthy individuals, validation in disease states is limited. Despite these caveats, it is not difficult to detect protein depletion when it is specifically looked for.

PROTEIN TURNOVER

It is, however, not only the amount of protein present in the body that is important, but whether it is able to function normally. The body's protein is in a dynamic state; proteins are continually being broken down to their constituent amino acids and resynthesized (Figure 1). There are two major advantages to protein turnover. Firstly, a dynamic protein turnover makes the maximum use from a minimum amount of amino acids. The body has no spare amino acid stores as it has for carbohydrate (glycogen) and fat (adipose tissue). Secondly, protein turnover is a major means of metabolic regulation. The disadvantages of a dynamic protein turnover is that it is very expensive in energy costs.

The more important a protein's role in the regulation of metabolism is, the faster it turns over. Thus, enzymes, particularly those at the branch points of metabolic pathways, have very short half lives whereas structural proteins such as collagen hardly turnover at all (Table 3; Figure 2). Other proteins with short half lives are those involved in immunologic surveillance.

Protein turnover is an example of a substrate cycle. A substrate cycle exists when opposing, non-equilibrium reactions catalyzed by different enzymes are active simultaneously. Metabolically important cycles have been proposed for all three macronutrients. For lipids, there is the breakdown of stored fat (lipolysis) and the resynthesis (reesterification) of triglycerides (triglyceride-fatty acid cycling). The major carbohydrate cycle is the conversion of glucose to glucose-6-P and back to glucose[2,59,76,77] Unlike the carbohydrate and lipid

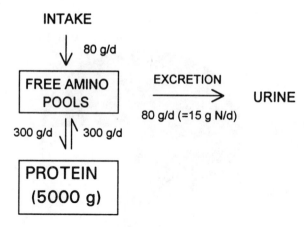

Figure 1. Relationship of protein intake to total protein turnover in a 70 kg man. The total protein pool is about 5000 g. About 300 g of new protein are made per day. Dietary intake provides only about 80 g of amino acids per day.

TABLE 3.
Some Representative Protein Turnover Rates.
Enzymes Have the Fastest Protein Turnover Rates
and Structural Proteins the Slowest

Protein	Tissue	~Half Life
Ornithine decarboxylase	liver	0.2 (hr)
Tyrosine aminotransferase	liver	2.0 (hr)
Phosphoenolpyruvate carboxylase	liver	5.0 (hr)
Glucokinase	liver	12.0 (hr)
Glucose-6-phosphate dehydrogenase	liver	15.0 (hr)
Fructose-bisphosphate aldolase	muscle	120 (hr)
Lactate dehydrogenase	muscle	144 (hr)
Cytochrome C	muscle	144 (hr)
Myosin	muscle	weeks
Elastin	various	months
Collagen	various	years

cycles mentioned above, protein turnover involves many steps in both the forward and backward directions. Because many of the steps, especially for protein synthesis, require ATP, protein turnover is the most energetically expensive of the macronutrient cycles.[2,59,73,76]

The importance of turnover can be illustrated with an example. Suppose that the liver has an acute need to double the rate of production of a particular protein. Three situations can be envisaged. Situation 1, with no cycle involved, would be to activate the necessary mRNA/protein synthesis machinery. Having

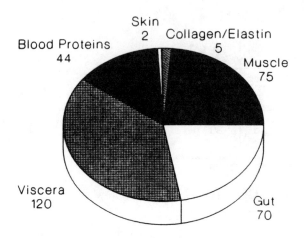

Figure 2. Amount of protein made per day by different tissues.

to start from the beginning is not likely to result in a rapid response. Situations 2 and 3 assume the presence of an active substrate cycle. Assume that in the basal state the rate of protein synthesis is 10 mg/min and of protein breakdown 9 mg/min. Net throughput (flux) is the same as without the cycle. If the rate of the forward reaction (protein synthesis) is increased by only 10% the through-put doubles (situation 2). In situation 3, the back reaction (protein breakdown) rate is decreased by 10% and again the net product flux is increased almost twofold. Without cycling, a 10% increase in the forward reaction rate leads to a 10% increase in product production. With substrate cycling, a 10% increase in either synthesis or breakdown rates leads to doubling of the rate of protein accumulation (Table 4).

Thus small changes in hormones, activators or other regulators can rapidly effect large change in protein concentration. This ability to respond rapidly allows the body to maintain levels of many different proteins and to rapidly increase the concentration of any protein as needed. Protein turnover is also necessary because there is no space in the body to maintain optimal levels of all proteins at their optimal concentrations. Many proteins are needed at high concentrations only intermittently; for example, immunoglobulins and host defense proteins are needed to combat external threats which can be infrequent and unpredictable.

Because there are no 'reserve' amino acids, any loss of protein means that somewhere in the body there will be a shortfall of amino acids for making protein. Initially the deficit is met by skeletal muscle. Skeletal muscle is by far the largest metabolically active protein pool in the body (collagen turnover is extremely slow [<1 year] and collagen does not contain most of the amino acids). The consequence is a progressive loss of muscle mass. Eventually a point is

TABLE 4.
Importance of Protein Turnover (Substrate Cycling) in Regulating Tissue Protein Levels

	Enzyme Activities		Throughput (net flux)	% Increase in flux
	Forwards	Backwards		
Situation 1 (no cycling)				
Basal	1	0	1.0	
10% increase in PSR	1.1	0	1.1	10
Situation 2 (with cycling)				
Basal	10	9.0	1.0	
10% increase in PSR	11	9.0	2.0	200
Situation 3 (cycling)				
Basal	10	9.0	1.0	
10% decrease in PBR	10	8.1	1.9	190

Note: Abbreviations: PSR, protein synthesis rate; PBR, protein breakdown rate. The table shows how cycling amplifies the effect of a 10% increase in protein synthesis or a 10% decrease in the breakdown rate on protein accumulation. Situation 1 is with no cycling; the pathway from starting material A to product P is simple and direct. Situations 2 and 3 involve cycling and differ in whether the forward or back reaction is regulated.

reached when other tissues start to lose amino acids as the body restricts progressively protein synthesis to the more essential functions. Eventually the low priority protein pools can be depleted no further and the more important protein pools become depleted. And then, if the body is challenged by an infection, the normal response mechanisms such as mobilization of the immune system are compromised because the available resources are limited, and in mobilizing the immune system, another protein pool becomes weakened and vulnerable, for example the lungs. It is a case of 'robbing Peter to pay Paul'. This is why the immediate cause of death from starvation is actually pneumonia.

The same result can occur with either an energy or protein deficiency. With the former, it is the amino acids that have become limiting; with the latter, the energy is not available. The protein deficiency situation is the more serious because the resources just are not there. With an energy deficiency the body can partially adapt by conserving energy, for example, by reducing the rate of protein turnover. But if mobilization is needed, it can be effected. It will take longer because total protein turnover will have been reduced.

The protein turnover theory explains why declining nutritional status results in a vicious cycle of continued depletion of lean body mass, weakening of host defenses leading to more and more prolonged infections, leading to further weakening of the body. It is important to point out the deficiencies are in the macronutrients, protein and energy, and not vitamins and minerals; so, assessing a patient's nutritional status should be focused on energy and protein metabolism.

PROTEIN STATUS OF AIDS PATIENTS

The most obvious sign of protein wasting in AIDS patients is the loss of skeletal muscle. When protein wasting is assessed by anthropometric measurements, it is muscle protein that is actually measured. It is then assumed that since muscle protein is depleted, other protein systems are similarly depleted. In AIDS patients the protein wasting appears to occur with both an inadequate and an apparently adequate dietary intake. An example of each situation is described below.

Malnutrition with Clearly Inadequate Intake

Trujillo et al. assessed the incidence of malnutrition among patients seen at a large urban medical center.[74] Over a 12-month period, 61 consecutive patients were seen and were given a detailed nutritional assessment. Some of the findings are tabulated in Table 5.

Notice the severe deficiency in energy and protein intake. Irrespective of the presence of any active disease process, this degree of reduced intake (~30%) will inevitably lead to chronic weight loss. By any standards, these patients are malnourished; apart from the weight loss, their serum albumin levels are low; 42% were moderately hypoalbuminemic. Further investigation of the 45 patients for whom a complete nutritional data set was available showed that 10 were marasmic, 5 hypoalbuminemic, 20 showed a mixed profile of protein and energy malnutrition and only 7 (~20%) could be considered to be in a good nutritional state.

Malnutrition with an Apparently 'Adequate' Intake

AIDS patients can be malnourished even if at the time of study energy intake is in the normal range. Kotler compared five clinically stable patients with AIDS, six seropositive controls, and five seropositive heterosexual control subjects.[46] The probable difference between this study and that reported by Trujillo is the small number of (selected?) patients studied by Kotler, the fact that they were clinically stable, whereas those studied by Trujillo were all the patients seen over a period of a year. Irrespective of intake, both Tables 1 and 2 show that the AIDS patients studied were malnourished. A most important feature of the disease is that although the patients are clearly undernourished, they do not eat more to compensate. It should also be pointed out that the effects of malnutrition develop over an extended time period; a nutritional recall or balance study conducted over a 1-week period may not be representative of the preceding months or years.

Whatever the cause of malnutrition, it is essential to be able to easily identify those patients who are malnourished so that treatment can be given. The best generally applicable marker for undernutrition is weight loss, although sometimes weight loss can be masked by edema. In one study, Singer

TABLE 5.
Metabolic Data on Weight Stable,
Asymptomatic AIDS Patients

	Controls	AIDS
PSR (g·prot·/kg/d)	4.38	3.16*
PBR (g·prot/kg/d)	4.66	3.57*
Fibgn. Syn. (%/d)	21.8	9.90*
Glucose Cycling (μM/kg/min)	9.70	7.15*
Total AA (μM/L)	3710	3142*
EAA:NEAA (%)	61.2	54.1*

Note: Abbreviations: PSR, whole body protein synthesis; PBR, whole body protein breakdown; Fibgn. Syn., fibrinogen fractional protein synthesis rate; AA, amino acids; EAA, essential amino acids; and NEAA, non-essential amino acids. Glucose cycling is the glucose substrate cycling through glucose-6-phosphate and back to glucose.

* $p < 0.05$.

Adapted from Stein T.P., et al., *Metabolism,* 39, 876, 1990.

et al. reported while there was a decrease in the total body potassium and extracellular water, there was a concomitant increase in the intracellular water.[66] This relative expansion of the total body water pool minimizes the (observable) total body weight loss. Thus measurement of body mass alone can result in under-estimation of the actual extent of any changes in lean body mass.[66]

Serum albumin is a useful marker for undernutrition in hospitalized patients, but its use for AIDS patients is not as clear-cut as with other wasting diseases. Chlebowski et al.[13] showed that patients with normal serum albumin levels lived longer than patients with low serum albumin levels. Singer et al. found that transferrin and retinol binding protein were better indicators of malnutrition.[66] The problem with albumin is that while a low value indicates either malnutrition or edema, a normal value does not necessarily indicate good nutritional status because the plasma albumin levels are dependent on fluid distribution within the body. Dehydration can result in anomalously high plasma albumin values.[66]

The protein wasting caused by AIDS is the same as that from simple starvation and other wasting diseases.[27,44,46,48] The changes in body composition with weight loss in AIDS patients measured by Singer et al. by bioimpedance were the same as those found in undernourished patients without AIDS.[66] The relationship between nutritional depletion and death in AIDS patients is the same as that for simple uncomplicated starvation.[44] Kotler et al. followed a group of AIDS patients followed until they died.[44] Whole body potassium, which is a good indicator of lean body mass because most of the potassium in the body is intracellular, declined steadily as did body fat. Not surprisingly, patients with diarrhea fared worse because their effective nutritional intake was lowest. And, as expected for malnutrition, the timing of death correlated with the depletion of LBM (~54%) and weight loss (~66% of the ideal body weight) rather than any specific disease.

Even though patients may appear to be adequately nourished, they can still be metabolically malnourished with the characteristic reduction in protein turnover.[75] Stein et al. investigated the metabolic status of ambulatory, asymptomatic, weight stable AIDS patients.[72] The patients were similar to those described by Kotler et al. (Table 5), being asymptomatic at the time of study. The objectives of the study were to determine whether (1) during periods of remission the disease process was quiescent and (2) if not, was there any evidence for ongoing hypermetabolism in the absence of any active infection. The study involved the measurement of the whole body protein synthesis, breakdown and fibrinogen synthesis rates and glucose cycling. All of these parameters are increased with hypermetabolism and decreased with undernutrition.[9,37,71,73,76,77]

The results of the study showed that the patients were malnourished rather than hypermetabolic (Table 5). Thus protein turnover, fibrinogen synthesis and glucose cycling were markedly lower in the AIDS patients. Because skeletal muscle is such a major contributor to protein turnover, it can be inferred that muscle protein turnover is reduced in these patients. This is to be expected since muscle protein turnover is invariably reduced with an energy deficit.[62]

In addition, the plasma aminogram showed decreased total amino acids and a reduced ratio of essential to non-essential amino acids in the AIDS group. An interesting side finding was that fibrinogen synthesis, a major acute phase protein was not decreased as much in patients treated with AZT than those given other treatments or no treatment (Figure 3). Fibrinogen is made in the liver and the disease does affect the liver, so, the better fibrinogen synthesis may be indicative of improved liver metabolism in the AZT-treated patients.

A potentially important consequence of the decreased whole body and acute phase protein synthesis would be a decreased ability to respond to subsequent challenges, and this could be an important factor in the increased susceptibility of AIDS patients to new infections. Although it is unlikely in AIDS patients that depressed whole body protein and fibrinogen synthesis and glucose cycling are the primary cause of immunodeficiency, they may be important facilitating factors contributing to the decreased ability of the patient to respond effectively to opportunistic infections.

The extreme weight loss associated with the later stages of AIDS and ARC is obviously a major factor affecting survival.[31,43,44,53,54,60] Whether inadequate intake is the primary cause or a consequence of the disease related metabolic changes is not known. The phrase 'inadequate intake' has mechanistic implications. All it means is that utilization is in excess of intake. A more recent study by Singer reported that for some of the patients, the magnitude of the

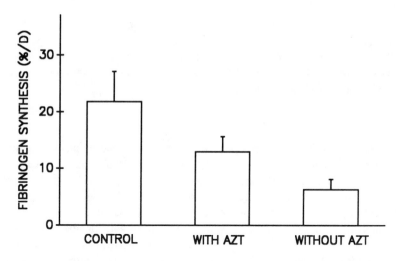

Figure 3. Effect of AIDS and AZT on fibrinogen protein synthesis rate. (Adapted from Stein, R.P. et al., *Metabolism*. 39, 876, 1990.)

weight loss is far greater than to be expected from the reduction in food intake alone.[66] Inadequate nutritional intake occurs with anorexia, malabsorption,[15,18,24,26,56] presence of malignancies in the gastrointestinal tract,[28,29] hypermetabolism[9,37] and a combination of these factors.

MECHANISMS OF WEIGHT LOSS

Muscle protein loss due to metabolic effects (as opposed to neuromuscular) can occur via one of four general mechanisms, inadequate intake, malabsorption, altered metabolism and atrophy secondary to disuse. At the tissue level inadequate intake and malabsorption are the same, a deficit in nutrient availability. In the one case it is due to decreased intake, in the other to a failure to adequately absorb nutrients from the gut. Hypermetabolism is different; it can result in weight loss even when nutritional intake is apparently adequate.[9,37] But irrespective of the mechanism, initially the shortfall in dietary protein is made up by the mobilization of protein from skeletal muscle. Preventing this loss of muscle protein from occurring requires that the underlying mechanism be known. For inadequate intake, increased nutrition is indicated, for hypermetabolism, pharmacological treatment of the cause of the hypermetabolism and for malabsorption, either pharmacological, therebyl or providing nutrition by a route other than via the gut, and for disuse atrophy, exercise.

For AIDS patients, the etiology of the body protein loss is clearly multifactorial with anorexia, malabsorption and hypermetabolism certainly being involved and maybe muscle, with some atrophy secondary to decreased activity. The relative importance of these four processes and whether this changes with progression of the disease is not known. Superimposed on these four direct causes are the consequences of any opportunistic infections, especially if they involve the gastrointestinal tract. Antiviral therapy can also result in decreased intake and so can psychosocial factors such as depression or embarking on an imbalanced 'fad' diet.

Undernutrition

Undernutrition with no obvious cause such as malabsorption, obstruction, GI distress or pain associated with eating is a complex phenomenon. Causes range from anorexia nervosa and depression (psychological), unavailability of food (economic) to decreased appetite as a side effect of chemotherapy (physiological), but in general there are no defined metabolic derangements which can be targeted for treatment. The cachexia associated with metabolic diseases, such as cancer, differs from the weight loss found with simple anorexia. Anorexia leads to starvation due to reduced intake; normal energy-conserving mechanisms are operative, whereas with metabolic disease, energy expenditure and protein turnover are increased and the normal adaptive responses to decreased intake impaired.[9,37]

Food Intake

A 1985 study by Chlebowski, using 3-day food records, reported that intake was inadequate in AIDS and ARC patients.[12] As the awareness of the association of undernutrition and its consequences became known in the AIDS community, AIDS patients made a conscientious effort to maintain intake. It does appear that reports of depressed food intake are more prevalent in the earlier rather than the more recent literature. Thus, a more recent review by Hellerstein concluded that dietary intake was adequate in most HIV-infected persons.[33]

Nevertheless, poor dietary intake persists when there are underlying pathological complications. An analysis of the reasons for undernutrition in a group of 50 patients by Ysseldyke reported 88% to be malnourished as defined by weight loss >10% and/or a serum albumin value less than 30 g/l.[81] Fever was present for 60%, anorexia 50%, nausea and vomiting in 20% and dysphagia in 10%. It is significant that this high degree of nutritional inadequacy was found despite the use of nutritional support for 70% of the patients.

Malabsorption

Malabsorption and diarrhea are frequent findings in AIDS patients and this in itself can often account for weight loss.[13,24,29] As the disease progresses, a wide range of gastrointestinal pathologies develop, ranging from neoplasms to a variety of infections to a diarrhea-producing malabsorption syndrome. It appears that as many as 50% of the patients with AIDS have gastrointestinal involvement. Malabsorption is common in patients with diarrhea as demonstrated by the D Xylose and ^{14}C tripalmitin and triolein absorption tests.[24,42,46]

In cases of malabsorption, intake is often in the normal range. This is illustrated in the data below, taken from a study of energy balance and body composition by Kotler of five clinically stable patients with AIDS, six seropositive controls, and five seropositive heterosexual control subjects[46] (Table 6). Body cell mass was significantly less in the AIDS group. Food intake was apparently 'normal' i.e. not different from the controls, but the resting energy expenditure was low. Intestinal absorption of carbohydrate xylose and a long chain triglyceride, triolein, were both significantly diminished leading to decreased availability of food energy.[46]

Others have shown that even in patients free of clinical malabsorption symptoms the potential for malabsorption exists. In one study, duodenal biopsies demonstrated chronic inflammation in 4 of the 10 patients studied.[24] Some authors have even suggested that the intestinal manifestations were a direct consequence of the virus.[18,24,26,58] However, a study by Smith et al. showed that well-known enteric pathogens could be identified in 85% (17 of the 20) of the patients investigated.[68] Almost certainly the disease predisposes the gastrointestinal tract to infection even if the GI tract is not the primary site of infection.

Altered Metabolism

Hypermetabolism

Most authors now consider the metabolic response to infection, trauma or presence of a tumor to be a manifestation of a common response pattern involving lymphokines and other macrophage-originated factors such as TNF.[9,57] Thus, the same pattern of metabolic abnormalities in host intermediary metabolism found with infection, trauma and sepsis occurs with cancer cachexia.[5,14,31,37,39,52]

Characteristic features of the hypermetabolic response are increased protein turnover, acute phase protein synthesis, gluconeogenesis, lipolysis, increased substrate cycling and loss of body protein.[9,39,41,62,71,76] There is also loss of body protein. The energy component is provided for by the mobilization of endogenous nutrients from peripheral muscle and fat.[9,39,52] A particularly interesting feature of the hypermetabolic response is that endogenous energy is still mobilized even when exogenous intake is apparently adequate.[9,39,59] The loss of body protein can be attenuated, but not prevented by nutritional means.[9,39]

As long as there is hypermetabolism, the loss of body protein will continue. The cancer patient is probably a better analogy than the injured or septic patient. Protein metabolism in cancer patients shows a chronic, low grade hypermetabolic response. The whole body protein synthesis rate is significantly elevated, albeit by a small amount, over control values. Over an extended period of time this will inevitably result in the chronic loss of body protein unless the cancer is treated. Although the death from cancer is usually

TABLE 6.
Malabsorption of Lipid (Triolein) and Carbohydrate in
AIDS Patients

	AIDS	Homosexuals	Heterosexuals
Xylose (urine)	22 ± 5*	32 ± 7	47 ± 5
Triolein (curve area)	3.75 ± 1.29*	6.16 ± 0.86	10.36 ± 2.51

* $p < 0.05$.

Adapted from Kotler, D.P., et al., *Am. J. Clin. Nutr.*, 51, 7, 1990.

not directly attributed to protein depletion, in many cases it was the cause. As the body's protein metabolism becomes progressively weakened, the ability to respond to stresses and opportunistic infections progressively diminishes and eventually the patient succumbs.

Evidence for Hypermetabolism in AIDS Patients

In the context of the AIDS patients, 'metabolic derangements' is probably a better term than 'hypermetabolism'. In the literature the two terms are used interchangeably. There is some evidence that hypermetabolism occurs in AIDS patients, and that the hypermetabolism is not an indirect consequence of a related disease process such as infection, but rather as a primary response to the HIV infection. As pointed out above, the response to negative energy balance by AIDS patients is different from that found with simple starvation. In uncomplicated starvation, fat is mobilized and lean body mass preserved. The opposite occurs in AIDS patients; there is loss of lean body mass while body fat is preserved. And, in the more extreme cases of hypermetabolism such as is found in severely injured and septic patients, the loss of lean body mass cannot be prevented by even very aggressive nutrition.[6,39]

A number of authors have suggested that hypermetabolism might be a factor in the (unexplained) weight loss in some AIDS patients.[11,31,26,34,35] A key indicator of hypermetabolism is increased energy expenditure. The question of whether energy expenditure is increased is very important for two reasons. Firstly, it is diagnostic for hypermetabolism, and secondly increased energy expenditure without a matching increase in usable energy intake is invariably fatal because it leads to progressive body wasting. However, the whole body energy expenditure is very dependent on body composition. Only lean body mass contributes to the whole body energy expenditure; so, fat depletion will increase the energy expenditure when expressed as per kg body weight. Thus normalizing energy expenditure to body weight can be misleading. The problem can be avoided by studying non-depleted patients or, preferably, actually measuring lean body mass and expressing the data as per kg lean body mass.

Several independent studies have now shown that HIV infection is associated with increased resting energy expenditure rates. The increased energy expenditure is found early in the course of HIV infection and increases further with AIDS.[30,34,35,53,66] Table 7 shows the data from one such report. Hommes et al. measured dietary intake, body composition and the BMR of 14 HIV-seropositive men with slow weight loss and 11 healthy controls.[35] All patients were clinically stable, free of acute infectious illness for at least two months before the study and without evidence of malabsorption or gross undernutrition. CD4+ lymphocytes were normal, suggesting that the host is responding to the HIV infection even though outwardly the virus appears to be latent.[35] The increased energy expenditure is also found after nutritional repletion with TPN when body composition should be normal.

The exact mechanism whereby hypermetabolism may cause an increase in energy expenditure and loss of lean body mass is not known. One likely possibility is that host defense mechanisms, such as increased acute phase protein synthesis, and substrate cycling, are increased and these result in increased energy expenditure. The real problem may be that dietary intake is not increased correspondingly. Increased substrate cycling may be responsible for much of the increased energy expenditure found with hypermetabolic states.[76] A net energy shortage can lead to protein loss because of lack of energy to maintain protein synthesis.

Muscle as a Source of Glutamine

In severe hypermetabolic states there is increased release of glutamine by skeletal muscle.[59,63,79] Whether this is a integral part of all hypermetabolic responses is not known. To date increased glutamine release has not been shown to occur with chronic low grade hypermetabolic states such as cancer; so, in the discussion below, increased glutamine release by skeletal muscle is treated as a separate entity from hypermetabolism.

Recently it has become apparent that muscle has a very important role in the immune system.[59] During times of stress there is increased activity by the lymphocytes, macrophages and other cells involved in the immune system. Glutamine is a preferred fuel for these cells. Glutamine can provide both an energy source and the precursors for new cell synthesis (nucleic acids, complex carbohydrates etc.) Although the needs could be met by glucose, glucose is spared because the glucose is needed for other tissues and the endogenous supply of glucose is very limited. The preferential use of glutamine by the cells involved in the immune system spares the limited glucose supplies to be available for those tissues that are obligate glucose users (brain, red blood cells).

The source of the glutamine is the breakdown of skeletal muscle.[59,79] Most of the amino acids derived from protein breakdown that are not reincorporated into protein are converted to alanine and glutamine. It is therefore possible that a chronic factor to the muscle protein loss in AIDS patients is a chronic need to supply glutamine to support the immune system. This is now believed to be

TABLE 7.
Energy Expenditure of HIV-Infected Men Free of Clinically Active Opportunistic Infections

	Patients	Controls
Weight (kg)	68.2	78.1*
Fat free mass (kg)	55.7	62.9*
Fat free mass (%)	81.6	80.7
RER (kcal/kg/d)	25.8	21.6*
RER (kcal/FFM/d)	31.6	26.8*
Energy intake (kcal/kg/d)	30.7	30.6

* $p < 0.05$.

Adapted from Hommes, M.J.T., *Metabolism,* 39, 1186, 1990.

the case in severe stress states such as sepsis. In both sepsis and trauma, giving supplemental glutamine decreases the net muscle protein loss and improves muscle protein synthesis.[79]

A primary user of the glutamine released by muscle is the gut.[3,59,80] The gut is a major factor in the immune system; about 25% of the gut cells are immune cells[1] and the gut is severely impacted by AIDS. It is therefore plausible that based on studies in other systems where both the gut and immune systems are compromised, such as sepsis, that providing supplemental glutamine to AIDS patients may simultaneously decrease muscle protein loss and improve immune status and alleviate malabsorption.[7,32,67,69,80]

Disuse Atrophy

Disuse of muscle inevitably results in atrophy of the muscle and this may be a factor in the muscle wasting found with AIDS patients. It is obviously possible that AIDS patients are less active, either because of the disease or for psychogenic reason or indirectly because of a change in lifestyle and this decreased activity will result in muscle loss. Exercise stimulates muscle development. Two studies have now shown that an exercise program increases physical fitness.[50,70] The study by Spence et al. clearly showed that exercise results in improvement of muscle function and body weight in AIDS patients.[70]

Spence et al. studied the effect of progressive resistance exercise (PRE) on body weight, muscle functional capacity and muscle protein content (estimated by anthropometry) in 24 males who were post therapy for acute *pneumocystis carinii* and pneumonia. The subjects were randomly assigned into two groups, an experimental group who engaged in PRE three times a week for six weeks and a control group who did not do any additional exercise. Some of the results are shown in Table 8. The control group continued to lose body weight and muscle mass (as indicated by the sum of the skinfold measurements), and

TABLE 8.
Effect of Exercise Therapy

Variable	Controls		AIDS	
	Prestudy	Post Study	Prestudy	Post Study
Weight (kg)	69.3	67.4*	70.8	72.5*
SSI (mm)	61.9	57.0*	65.4	70.5*
Flexiom (J)	773	688*	854	1059*

Note: Abbreviations: SSI, sum of skinfolds; Flexion total work in Joules.

* $p < 0.05$ vs prestudy period.

Adapted from Spence, D.W., et al., *Arch. Phys. Med. Rehab.*, 71, 644, 1990.

muscle functional ability decreased with time. In contrast, in the PRE group, there was weight gain, increased muscle mass and improved muscle function. A probable contributing factor to the weight gain is that the exercise regime stimulated the appetite and this led to an increased intake of food.

However, there is some evidence that chronic exercise can lead to immune suppression.[38] Literature on this topic is conflicting with some reports indicating immunosuppression[8,61] and others, stimulation.[19,22,23,51,64] Clearly it would be counterproductive if exercise improved the status of muscle but compromised the immune system. Rigsby et al. investigated this question by studying the effects of chronic exercise on aerobic capacity and strength and immune status in 37 HIV-positive men with modified Walter Reed scores ranging from 1 to 5.[65] Immune status was determined by the total leukocyte, lymphocyte, CD4+, CD8+ and the CD4+:CD8+ ratio. As with the study of Spence et al., there was a marked improvement in fitness. No statistically significant changes in any of the immune parameters assayed were found. But for each parameter measured there was a trend for improvement rather than impairment. Taken together it would seem that in this study there was a benefit to the immune system from the exercise and that had a larger group of subjects been studied, significance might have been found.

ACKNOWLEDGMENT

This work was supported by NIH grant #DK-41927.

REFERENCES

1. Alverdy JC. Effects of glutamine supplemented diets on immunology of the gut. *J.P.E.N.* 14: suppl. 109S-113S, 1990.
2. Ardawi MS and Newsholme EA. *Essays Biochem.* 21:1-44, 1985.

3. Ardawi MS, Jamal YS, Ashy AA, Nasr H and Newsholme EA. Glucose and glutamine metabolism in the small intestine of septic rats, *J. Lab. Clin. Med.* 115:660-668, 1990.

4. Askenazi JA, Weissman C, Rosenbum SA, et al. Nutrition and the respiratory system. *Crit. Care. Med.* 10:163-187, 1982.

5. Bennegard K, Lindmark L, Eden E, et al. Flux of amino acids across the leg in weight-losing cancer patients. *Cancer Res.* 44:386-393, 1984.

6. Brendan MF. Uncomplicated starvation versus cancer cachexia. *Cancer Res.* 37:2359-2364, 1977.

7. Burke DJ, Alverdy JC, Aoys E and Moss GS. Glutamine-supplemented total parenteral nutrition improves gut immune function. *Arch Surg.* 124:1396-1399, 1989.

8. Cannon JG and Klugir R. Endogenous pyrogen activity in human plasma after exercise. *Science.* 220:617-618, 1983.

9. Cerra FB. Hypermetabolism, organ failure and metabolic support. *Surgery.* 101:1-14, 1987.

10. Chandra S and Chandra RK. Nutrition, immune response and outcome. *Prog. Food. Nutr. Sci.* 10:1-65, 1966.

11. Chlebowski RT. Significance of altered nutritional status in acquired immune deficiency syndrome (AIDS), *Nutrition and Cancer.* 7: Nos. 1 & 2, 85-91, 1985.

12. Chlebowski RT, Grosvenor M, Kruger S, et al. Dietary intake in HIV infection. Relative caloric deficiency in patients with AIDS. 5th International AIDS Conference, Montreal, Canada, page 467 (Abstr.), June, 1989.

13. Chlebowski RT, Grosvenor MB, Bernhard NH, et al. Nutritional status, gastrointestinal dysfunction, and survival in patients with AIDS. *Am. J. Gastroenterol.* 84:1288-1293, 1989.

14. Clark ER, Lewis AM and Waterhouse C. Peripheral amino acid levels in patients with cancer. *Cancer.* 42:2909-2913, 1978.

15. Colman N and Grossman F. Nutritional factors in epidemic Kaposi's sarcoma, *Sem. Oncol.* 14:54-62, 1987.

16. Cornlath DR, McArthur JC, Kennedy PG, et al. Inflammatory demyelinating peripheral neuropathies associated with human T-cell lymphotrophic virus type III infection. *Ann. Neurol.* 21:32-40, 1987.

17. Cunningham-Rundles S. Effects of nutritional status on immunologic function. *Am. J. Clin. Nutr.* 35:1202-1210, 1982.

18. Dworkin B, Wormser GP, Rosenthal WS, et al. Gastrointestinal manifestations of the Acquired Immunodeficiency Syndrome: A review of 22 cases. *Am J of Gastroenterol.* 80, 774-778, 1985.

19. Edlund A. The effect of defined exercise on the early convalescence of viral hepatitis. *Scand. J. Infect. Dis.* 3:189-196, 1971.

20. Eidelberg D, Sotrel A, Vogel H, et al. Progressive polyradiculopathy in acquired immunodeficiency syndrome. *Neurology.* 36:912-916, 1986.

21. Elder G, Dalakis, Pezeshkpour G and Sever J. Ataxic neuropathy due to ganglioneurotonits after probable acute human immunodeficiency virus infection. *Lancet.* 2:1275-1276, 1986.

22. Friman G, Ilback N, Beisel WR, et al. The effect of strenuous exercise on infection with Franciscella tularrensis in rats. *J. Infect. Dis.* 145:706-714, 1982.

23. Gatmaitan BG, Chason JL and Lerner AM. Augmentation of the virulence of immune coxsackie-virus B-3 myocardiopathy by exercise. *J. Exp. Med.* 131:1121-1138, 1970.

24. Gillin JS, Shike M, Alcock N, Urmacher C, et al. Malabsorption and mucosal abnormalities of the small intestine in the Acquired Immunodeficiency Syndrome, *Ann. Intern. Med.* 102:619-622, 1985.

25. Gleason WA and Roodman ST. Reversible T cell depression in malnourished infants with pneumocystis pneumonia. *J. Pediatr.* 90:1023-1033, 1977.

26. Gottlieb MS. The Acquired Immunodeficiency Syndrome, *Ann. Intern. Med.* 99:208-220, 1983.

27. Gray RH. Similarities between AIDS and protein-calorie malnutrition. *Am. J. Public Health.* 73:1332, 1983.

28. Greenson JK, Belistos PC, Yardley JH, et al. AIDS enteropathy: occult enteric infections and duodenal alterations in chronic diarrhea. *Ann. Intern. Med.* 114:366-372, 1991.

29. Groopman JE. Neoplasms in the Acquired Immune Deficiency Syndrome: The multidisciplinary approach to treatment, *Sem. Oncol.* 14, No 2 1-6, 1987.

30. Grunfeld C, Pang M, Shimuzu L, et al. Resting energy expenditure, caloric intake, and short term weight change in human immunodeficiency virus infection and the acquired immunodeficiency syndrome. *Am. J. Clin. Nutr.* 55:455-460, 1992.

31. Grunfeld C and Feingold KR. Metabolic disturbance and wasting in the acquired immunodeficiency syndrome. *New Engl. J. Med.* 327:329:337, 1992.

32. Hammarqvist F, Wernerman J, Von der Decken A and Vinnars E. Alanyl-glutamine counteracts the depletion of free glutamine and the postoperative decline in protein synthesis is skeletal muscle. *Ann. Surg.* 212:637-644, 1990.

33. Hellerstein MK, Kahn J, Mudie H, et al. Current approach to the treatment of human immunodeficiency virus associated weight loss: pathophysiologic considerations and emergent management strategies. *Sem. Oncol.* 17(suppl. 9):17-33, 1990.

34. Hommes MJT, Romjijn A, Endert E and Sauerwein HP. Resting energy expenditure and substrate oxidation in human immunodeficiency-virus related asymptomatic men: HIV affects host metabolism in the early asymptomatic stage. *Am. J. Clin. Nutr.* 54:311-315, 1991.

35. Hommes MJT, Romjijn A, Godfried, et al. Increased resting energy expenditure in human immunodeficiency virus infected men. *Metabolism.* 39:1186-1190, 1990.

36. Hughes WT, Price RA, Sisko F, et al. Protein calorie malnutrition: A host determination for pneumocystis carinii infection. *Am. J. Dis. Child.* 128:443-448, 1988.

37. Inculet RI, Stein TP, Peacock JL, Leskiw, MJ, Maher M, Gorschboth CM and Norton JA. Altered Leucine metabolism in non-cachectic sarcoma patients. *Cancer Res.* 47:4746-4749, 1977.

38. Keast D, Cameron K and Morton AR. Exercise and the immune response. *Sports Med.* 5:248-267, 1988.

39. Kern, KA and Norton JA. Cancer cachexia, *J. Parent. Ent. Nutr.* 12:286-298. No 3, 1988.

40. Keusch GT and Farthing MJG. Nutritional aspects of AIDS. *Annu. Rev. Nutr.* 10:475-501, 1990.

41. Kinney JM and Elwyn DE. Protein metabolism and injury. *Annu. Rev. Nutr.* 3:433-466, 1983.

42. Kotler DP, Gaetz, Lange M, et al. Enteropathy associated with the acquired immunodeficiency syndrome. *Ann. Intern. Med.* 101:421-8, 1984.

43. Kotler DP, Wang J and Pierson RN. Body composition studies in patients with the acquired immunodeficiency syndrome. *Am. J. Clin. Nutr.* 42:1255-1265, 1985.

44. Kotler DP, Tierney AR, Pierson RN, et al. Magnitude of body cell mass depletion and timing of death from wasting in AIDS. *Am. J. Clin. Nutr.* 50:444-447, 1989.

45. Kotler DP. Malnutrition in HIV infection and AIDS. *AIDS 1989.* 3(Suppl 1):S175-S180, 1989.

46. Kotler DP, Tierney AR, Brenner SK, et al. Preservation of short term energy balance in clinically stable patients with AIDS. *Am. J. Clin. Nutr.* 51:7-13, 1990.

47. Kotler DP, Tierney AR, Culpepper-Morgan JA, et al. Effect of home total parenteral nutrition on body composition in patients with acquired immunodeficiency syndrome. *J.P.E.N.* 14:454-458, 1990.

48. Kotler P, Tierney AR, Ferraro R, et al. Enteral alimentation and repletion of body cell mass in malnourished patients with acquired immunodeficiency syndrome. *Am. J. Clin. Nutr.* 53:149-154, 1991.

49. Kotler DP. Nutritional effects and support in the patient with acquired immunodeficiency syndrome. *J. Nutr.* 122 (Suppl):723-727, 1992.

50. LaPerriere AR, Antoni MH, Schneiderman, et al. Exercise intervention attenuates emotional distress and natural killer cell decrements following notification of positive serologic status for HIV-1. *Biofeedback and Self-Reguln.* 15:229-242, 1990.

51. Lewicki RH, Tchorzewski H, Denys A, et al. Effect of physical exercise on some parameters of immunity in conditioned sportsmen. *Int. J. Sports Med.* 8:309-314, 1987.

52. Lundholm K, Edstrom S, Karlberg I, et al. Glucose turnover, gluconeogenesis from glycerol, and estimation of net glucose cycling in cancer patients. *Cancer* 50:1142-50, 1983.

53. Malcolm JA and Sutherland DC. HIV and Nutrition. *Lancet* 338:645-649, 1991.

54. Malnutrition and weight loss in patients with AIDS. *Nutr. Rev.* 7:354-346, 1989.

55. Melchior J-C, Salmon D, Rigaud D, et al. Resting energy expenditure is increased in stable, malnourished HIV-infected patents. *Am. J. Clin. Nutr.* 53:437-441, 1991.

56. Miller TL, Orav EJ, Martin SR, et al. Malnutrition and carbohydrate malabsorption in children with vertically transmitted human immunodeficiency virus 1 infection. *Gastroenterology.* 100:12996-1302, 1991.

57. Moldawer LL, Georgieff M and Lundholm K. Interleukin 1, tumour necrosis factor-alpha (cachectin) and the pathogenesis of cancer cachexia, *Clin. Physiol.* 7, 263-274, 1987.

58. Nelson J, Wiley CA and Reynolds-Kohler C. Human Immunodeficiency virus detected from patients with gastrointestinal symptoms. *Lancet* 1:259, 1988.
59. Newsholme EA, Newsholme P, Curi, R, et al. A role for muscle in the immune system and its importance in surgery, trauma, sepsis and burns. *Nutrition*. 4:261-268, 1988.
60. Nutrition, immunomodulation and AIDS. Symposium, American Institute of Nutrition Annual Meeting, Atlanta, *J. Nutr.* 122:715-757, 1992.
61. Raskis HA. Systemic stress as an inhibitor of experimental tumors in Swiss mice. *Science*. 116:169-171, 1952.
62. Rennie MJ. Muscle protein turnover and the wasting due to injury and disease. *Br. Med. Bull.* 41:257-264, 1985
63. Rennie MJ, Maclennan PA, Hundal PS, et al. Skeletal muscle glutamine transport, intramuscular glutamine concentration and muscle protein turnover. *Metabolism*. 38 (Suppl. 1):47-51, 1989.
64. Reyes MP and Lerner AM. Interferon and neutralizing antibody in sera of exercised mice with coxsachie virus B-3 myocarditis. *Proc. Soc. Exp. Biol. Med.* 151:333-338, 1976.
65. Rigsby LW, Dishman RK and Jackson AW. Effect of exercise training on men seropositive for the immunodeficiency virus-1. *Med. Sci. Sports Exerc.* 24:6-12, 1992.
66. Singer P, Katz DP, Dillon L, et al. Nutritional aspects of the acquired immunodeficiency syndrome. *Am. J. Gastroenterol.* 87:265-273, 1992.
67. Smith RJ and Wilmore DW. Glutamine nutrition and requirements. *J.P.E.N.* 14(Suppl.):94S-99S, 1990.
68. Smith PD, Lane HC, Gill VJ, et al. Intestinal infections in patients with Acquired immunodeficiency syndrome (AIDS). Etiology and response to therapy. *Ann. Intern. Med.* 108:328-333, 1988.
69. Souba WW, Herskowitz K, Salloum RM, Chen MK and Austgen TR. Gut glutamine metabolism. *J.P.E.N.* 14(4 Suppl.):45S-50S, 1990.
70. Spence DW, Galantino MLA, Mossberg K and Zimmerman SO. Progressive resistance exercise: Effect on muscle function and anthropometry of a selected AIDS population. *Arch. Phys. Med. Rehab.* 71:644:648, 1990.
71. Stein, T.P., Leskiw, M.J., Oram-Smith, J.C. and Wallace, H.W. Changes in protein synthesis after trauma: importance of nutrition. *Am. J. Physiol.* 233:348-355, 1977.
72. Stein TP, Nutinsky C, Condolucci D, et al. Protein and energy substrate metabolism in AIDS patients. *Metabolism*. 39:876-881, 1990.
73. Stein TP, Rumpler WV, Leskiw MJ, et al. Effect of reduced calorie intake on protein turnover and glucose cycling in man. *Metabolism*. 40:478-453, 1991.
74. Trujillo WB, Borlase BC, Bell SJ, et al. Assessment of nutritional status, nutrient intake, and nutrition support in AIDS patients. *J. Am. Diet. Assoc.* 92:477-481, 1992.
75. Waterlow JC. Metabolic adaptation to low intakes of protein and energy. *Annu. Rev. Nutr.* 6:495-526, 1986.
76. Wolfe RR, Herndon DL, Jahoor F, Miyoshi H and Wolfe M. Effect of severe burn injury on substrate cycling by glucose and fatty acids. *N.E.J.M.* 317:403-407, 1987.

77. Wolfe RR. Radioactive and stable isotope tracers in biomedicine. *Principles and Practice of Kinetic Analysis*. New York, Wiley-Liss, 1992, Chapter 18, pp 377-382.
78. Wollschlager CM, Khan FA, Fahtkara RK and Shivaram R. Pulmonary manifestations of the Acquired Immunodeficiency Syndrome (AIDS), *Chest*. 85:No 2, 197-202, 1984.
79. Yoshida S, Lanza-Jacoby S and Stein TP. Leucine and glutamine metabolism in septic rats. *Biochem. J.* 276:405-409, 1991.
80. Yoshida S, Leskiw MJ, Schluter MD, et al. Effect of total parenteral nutrition, systemic sepsis and glutamine on gut mucosa in rats. *Am. J. Physiol.* 263:*(Endocrinol. Metab.)* E368-E373, 1992.
81. Ysseldyke ll. Nutritional consequences and incidence of malnutrition among AIDS patients. *J. Am. Diet. Assoc.* 91:217-218, 1991.

Chapter

12

AIDS and Food Safety

Ralph Meer

INTRODUCTION

Secondary infections contribute significantly to the morbidity and mortality of HIV-infected persons. The prevention of certain infections can be achieved by avoiding the consumption of contaminated food and water. Although a small number of known microorganisms are pathogenic to humans, certain pathogenic microorganisms are ubiquitous in nature and can easily become a contaminant of raw or prepared foods. In healthy individuals, the occurrence of foodborne disease is typically short-lived. However, in immunocompromised persons such infections may result in chronic gastroenteritis, septicemia, organ compromise, and potentially death. Foodborne illnesses can contribute to the compromised nutritional status of HIV-infected individuals which in turn can result in further compromised immunity. These illnesses can result in anorexia, malabsorption, and increased nutrient losses. Health professionals treating HIV-infected individuals need to be educated on the prevention of secondary infections in order to provide counseling to caretakers and patients on food safety principles.

FOODBORNE DISEASE

A variety of pathogenic, as well as nonpathogenic, microorganisms (i.e. viruses, fungi, bacteria, protozoa) may be transmitted via food.[1-55] Food may play an active role in disease transmission by supporting growth of the etiologic agent or a passive role where food does not support growth but services only as the vehicle of transmission.[1] There are two types of foodborne diseases: infections and intoxications. Foodborne infections result from eating food that

0-8493-7842-7/94/$0.00+$.50
© 1994 by CRC Press, Inc.

contains a pathogenic organism (i.e. agent ingested and multiplies in the gut), while foodborne intoxications are caused by eating food that contains a toxic chemical (i.e. agent has produced a toxin in food prior to ingestion). Foodborne disease typically presents itself as an enteric disease resulting in nausea, vomiting, and/or diarrhea with or without additional symptoms (e.g. fever, chills, headache, fatigue, etc.). Chronic diseases which may result from foodborne diseases include arthropathies, chronic gastroenteritis, organ compromise, nutritional and other malabsorptive disorders as well as the potential for death.[1]

ENTERIC INFECTIONS IN HIV-INFECTED INDIVIDUALS

One of the major contributors to the morbidity and mortality associated with HIV-infected persons is intestinal disease.[6] Diarrhea is usually the most significant manifestation of such an infection and a possible life-threatening complication.[7] Clinical observations suggest that decrease in immune function seen in HIV-infected patients results in a corresponding increase in enteric infections and other opportunistic processes[7,8] It has been demonstrated that the concentration of T lymphocytes in the intestinal mucosa mirrors that of the blood stream and could explain in part the predisposition to infections caused by enteric pathogens seen in HIV-infected patients.[9]

Many microorganisms (e.g. *Giardia lambia, Entamoeba histolytica, Cryptosporidium, Salmonella, Shigella, Listeria, Yersinia, Campylobacter,* Hepatitis A, etc.) identified as the cause of enteric infections in HIV-infected patients have also been recognized as etiological agents in food and waterborne disease.[1-5,7,10-12] Common organisms associated with food and waterborne diseases and their associated sources and symptoms are listed in Table 1.[13,14] Investigations have identified *Salmonella* infections as an important cause of morbidity in patients with HIV-infection.[15-20] Celum et al.[21] reported the incidence of *Salmonella* infections as being 20 times that of men without AIDS. The incidence of *Salmonella* bacteremia was 45% and 9% for patients with and without AIDS, respectively. It was suggested that the increased infection rate, as well as the unusual serotypes of *Salmonella* identified from AIDS patients, could be related to the use of "natural" food products by these individuals. Mascola et al.[22] reported on five cases of listeriosis in patients with AIDS between January 1985 and March 1986 in Los Angeles County. Since individuals with decreased cell-mediated immunity are the primary risk group for *Listeria monocytogenes* infections, it was recommended that HIV-infected persons avoid consuming food items associated with listeriosis (i.e. inadequately pasteurized milk and other dairy products as well as improperly washed raw fruits and vegetables). The identification of four HIV-infected patients with *Campylobacter jejuni* infections was described by Perlman and

TABLE 1.
Common Organisms Associated with Food and Waterborne Illness

Organism	Associate Foods	Onset/Symptoms
Salmonella (infection)	Unpasteurized milk, raw poultry, meat and eggs.	6–8 h. Nausea, stomach pain, diarrhea, fever, chills and headache.
Clostridium perfringens (intoxication)	"Cafeteria germ". Temperature abused meat and meat dishes, i.e. gravies, stews, dressing, casseroles.	9–15 h. Diarrhea, abdominal pain, gas sometimes nausea and vomiting.
Listeria (infection)	Raw milk, raw milk cheeses, seafood, leafy vegetables.	7–30 d. Usually 48–72 h. Fever, headache, nausea, vomiting. Complications: meningitis, septicemia, spontaneous abortion, stillbirth.
Campylobacter jejuni (infection)	Raw poultry, meat and unpasteurized milk.	2–5 d. Diarrhea, abdominal pain, and cramps.
Escherichia coli (infection)	Raw or rare ground beef and unpasteurized milk.	3–4 d. Hemorrhagic colitis, severe abdominal cramps, diarrhea (often bloody), nausea, vomiting and low grade fever.
Staphylococcus aureus (intoxication)	Meat, poultry and high protein dishes.	30 min to 8 h. Diarrhea, vomiting, nausea, abdominal pain and cramps.
Clostridium botulinum (intoxication)	Low acid canned foods such as meat, vegetables, especially home canned foods, Vacuum packed foods.	4–36 h. Neurotoxic symptoms: double vision, trouble speaking, swallowing, difficulty breathing.
Cryptosporidium (infection)	Contaminated water and potentially food.	2–14 d. Flu-like illness, chronic watery diarrhea, nausea, vomiting, abdominal cramping, fever, and malaise.
Entamoeba histolytica (amebiasis) (infection)	Polluted water and vegetables grown in contaminated soil.	3–10 d. Abdominal pain, cramps, diarrhea, weight loss, fatigue and anemia.
Giardia lamblia (Giardiasis) (infection)	Contaminated water or uncooked foods contaminated while grow- Contamination of foods by infected food handlers.	1–3 d. Explosive, watery stools, abdominal cramps, anorexia, nausea, vomiting.
Hepatitis A virus (infection)	Raw seafood from water contaminated by sewage. Can be spread via infected food handlers.	10–50 d. Malaise, anorexia, nausea, vomiting, diarrhea and jaundice.

Data from References 13, 14.

colleagues.[10] Three patients had persistent infections and one had bacteremia. It was demonstrated that the decreased ability to eradicate this organism from the intestinal tract resulted from a defective production of *Campylobacter* specific antibodies.

FOOD SAFETY COUNSELING

Knowledge of safe food-handling techniques is considered essential for HIV-infected patients and caretakers.[23] Archer[24] emphasized the need to incorporate counseling, for HIV-infected patients, covering the prevention of foodborne illness as a component of an overall strategy for defensive living. It was recommended that this be accomplished through a coordinated effort between federal agencies with food safety responsibilities and physicians treating HIV-infected patients. Griffin and Tauxe[25] recommended that immunocompromised patients receive counseling on food safety principles. Emphasis should be placed on the adequate cooking of animal foods (i.e. avoiding the consumption of raw or partially cooked animal foods) and cross-contamination of raw and cooked foods or foods not requiring cooking to help decrease risk of infection with enteric pathogens. Filice and Pomeroy[26] stressed the importance of educating patients on hidden or unrecognized sources of contamination including raw or undercooked eggs present in hollandaise sauce, meringue, caesar and other salad dressings, frosting, egg nog, ice cream, mayonnaise, and dough. It was also recommended that HIV-infected persons be instructed to avoid drinking untreated surface water or unsafe well water to help prevent secondary infections. A videotape covering proper food handling procedures is available for educational purposes.[27]

HOME CARE

Recommendations for home care sanitation to reduce foodborne illness in HIV-infected patients include: appropriate sanitation in the kitchen (e.g. counters, floors, utensils, food storage areas, etc.); proper personal hygiene of food preparers; avoiding consumption of raw or undercooked animal products; and the peeling of raw fruits and vegetables.[28] The separation of utensils used by HIV-infected patients is unnecessary. Thorough cleaning with hot soapy water, rinsing with hot water, and air drying or cleaning via the dishwasher using the hot water cycle is sufficient to control contamination.[28,29] A 1:10 dilution of household bleach can be used to disinfect surfaces that have been exposed to large amounts of blood or body excretions.[29,30]

The proper attention to food preparation is also paramount in the home care of children with HIV-infection.[31] Parents or other caretakers should be instructed in sanitary bottle and formula preparation. Putting infants or toddlers to bed with bottles is not recommended due to the possibility of bacterial growth in formula. Milk or juices held at room temperature for extended periods of time should be discarded. Children should not be spoon fed from the jar as this may allow for the inoculation of the food with oral flora. Unused portions of baby food in jars should be promptly refrigerated and consumed within 24 hours. As previously indicated for adults, children should not be fed animal foods that have not been adequately heat-treated, and raw vegetables and fresh fruits should be thoroughly washed prior to peeling or cooking. Table 2 summarizes proper sanitation and food safety precautions which should be observed by HIV-infected persons and caretakers to help prevent secondary infections.[6,23,32,33]

STERILE AND LOW MICROBIAL DIETS

The use of sterile and low microbial diets to provide "pathogen free" food has been studied for immunocompromised patients, particularly cancer and transplant patients; however, these studies have not included HIV-infected patients.[34] Without evidence that these diets are efficacious in HIV-infected patients and for practical reasons (e.g. difficulty preparing and maintaining sterile and low microbial conditions, as well as their questionable palability), these diets are not recommended for this patient population at this time.[35] Adherence to the previously listed food safety techniques will provide the greatest level of protection to patients with HIV disease.

ENTERAL FEEDINGS

HIV-infected persons may require enteral nutrition support to supplement diet intake or meet 100% of their nutritional needs. Enteral formulas are generally considered commercially sterile from the manufacturers; however, they have many opportunities, from opening to administration, to become contaminated. Most enteral formulas are considered an ideal growth medium for microorganisms.[36] A number of organisms including *Bacillus cereus,* B-hemolytic streptococci, *Enterobacter, Escherichia coli, Klebsiella, Moraxella, Proteus, Pseudomona aerginosa, Salmonella enteridis, Staphylococcus aureus, Staphylococcus epidermis,* and yeasts have been isolated from enteral formulas.[37-39] A study performed by Anderson and colleagues[40] demonstrated a significant association between the presence of diarrhea and the degree of bacterial contamination of the enteral formulas received by the patients. Also nonmanipulated formulas had significantly less contamination than locally prepared and manipulated formulas. An investigation by Freedland et al.[41]

TABLE 2.
Food Safety Recommendations

1.	Wash hands with soap and hot water before food handling; after using the restroom; after touching garbage or garbage containers; between handling raw and cooked foods; between handling dirty and clean dishes; and after eating, coughing, sneezing, or touching hair or face.
2.	Food preparation surfaces should be cleaned and disinfected frequently. Utensils and cutting boards used for raw food preparation must be changed or cleaned and disinfected before using them for cooked foods or foods not requiring cooking (e.g. raw vegetables and fruits).
3.	Avoid consumption of raw animal foods such as eggs, meat, poultry, fish, or other seafood or foods containing raw animal ingredients.
4.	Cook red meat (170°F) and poultry (180°F) until well done.
5.	Avoid cracked eggs.
6.	Use only pasteurized milk and other dairy products.
7.	Keep hot foods hot (cooked to 165° to 212°F and held at 140° to 165°F).
8.	Keep cold foods cold (refrigeration temperatures should be ≤40°F; freezers should be ≤0°F).
9.	Use leftover food products within 48 hours to avoid spoilage, reheat to ≥165°F.
10.	Do not allow perishable foods to stand at room temperature 45° to 140°F for more than 2 hours.
11.	Thaw frozen foods in refrigerator; submerged in cold water, changing water every 30 minutes; or in microwave; never at room temperature.
12.	Do not refreeze defrosted foods.
13.	Wash fruits and vegetables thoroughly followed by peeling when possible.
14.	Do not eat moldy or spoiled foods.
15.	Canned goods should not be used if the can is dented, swollen, damaged, or has no label.
16.	Do not use foods after the recommended expiration date on the label.
17.	When shopping, select perishable items last, wrap meats in plastic bags and return home with perishables as soon as possible.
18.	WHEN IN DOUBT, THROW IT OUT!

Data from References 6, 23, 32, 33.

demonstrated no relationship between the occurrence of liquid stools and degree of contamination of enteral formulas, although the presence of Gram negative bacilli was correlated with an increased incidence of abdominal distention.

The ability of microorganisms to survive and multiply in enteral feeding preparations requires strict quality control measures with respect to the handling of ingredients and the maintenance of stringent hygiene standards. Enteral formulas are typically administered at room temperature via continuous infusion over an 8–12 hour period. Recommendations for formula hang times range from 4–12 hours.[42-46] Recommendations to minimize contamination and growth of microorganisms in enteral formulas include: avoid using home-blenderized formulations; use commercially sterile diets only; use of sterile water in preparing formulas; rigorous equipment cleaning and sanitation; proper hand washing; use of a "no touch" technique when transferring liquid diets from original or preparation container to delivery container; aseptic technique in product preparation; limiting hang time to 4 hours; holding of opened formula (covered) at refrigeration temperatures; and the changing of delivery containers and tubing each 24 hours.[43,44,46-48]

FOREIGN TRAVEL

For non-HIV-infected persons, a case of travelers' diarrhea resulting from a foodborne or waterborne organism is usually mild, and at worst an inconvenience, lasting only a few days. However, in HIV-infected persons such an infection can potentially result in a life-threatening illness. Of utmost importance in the prevention against travelers' diarrhea is the careful selection of food and beverages.[49] The most common bacterial, viral, and parasitic causes of travelers' diarrhea are listed in Table 3.[49] Recommendations for avoiding food and waterborne transmission of travelers' diarrhea include: consuming only bottled water, soft drinks, or beer and wine in sealed containers; avoid tap water and ice cubes; purify drinking water by boiling or using iodine or chlorine preparations; avoid unpasteurized dairy products; consume well-cooked foods that are eaten while still hot; and avoid raw or rare meat, poultry, and fish, cold platters, custards, mayonnaise spreads, pastries, foods served at buffets, foods from street/roadside vendors, and fresh fruits and vegetables that cannot be peeled.[49-54]

FOOD SERVICE ISSUES

The situation for food service employers has been described as a "catch 22".[55] When faced with employees who are HIV-infected, balancing the rights of a worker with the concerns and fears of fellow employees and patrons can be difficult. The routine testing of food service workers for the presence of HIV antibodies is not recommended to prevent transmission of the disease from workers to customers.[56] Employment of HIV-infected individuals is not a violation of the Occupational Health and Safety Act since HIV disease is not transmitted by casual contact and does not pose a risk in the work place.[57] The Hatch Amendment to the American Disabilities Act of 1990 requires the identification of diseases that can be spread by food handlers by the Department of Health and Human Services. Since HIV infections are not believed to be transmitted in this manner, HIV-infected persons cannot be removed from food handling positions.[57]

TABLE 3.
Causes of Travelers' Diarrhea
Resulting from Contaminated
Food or Water

Bacteria
  Escherichia coli (toxigenic)
  Shigella
  Salmonella
  Campylobacter
  Aeromonas
  Plesiomonas
  Vibro (parahemolyticus, non-01, *cholerae)*
Viruses
  Norwalk agent
  rotavirus
Parasites
  Giardia lamblia
  Entamoeba hystolytica
  Cryptosporidium
  Isospora belli
  Blastocystis hominis
  Strongyloides stercoralis

From Hill, D. and Pearson, R., *Ann. Intern. Med.*,
108, 839-852, 1988. With permission.

It is recommended that all food service workers abide by the recommended standards and practices of good personal hygiene and food sanitation.[58] It is emphasized that all food service workers avoid hand injuries while preparing food, and if an injury should occur, both aesthetic and sanitary considerations would necessitate discarding food contaminated with blood.[56] Also, HIV-infected individuals do not need to be restricted from work unless they have an infection or illness that would exclude any other food service personnel from working in their position.[56]

Although the Centers for Disease Control has not documented the transmission of HIV infection via food or from the handling of trays, dishes, or eating utensils used by HIV-infected patients, to reduce fears of food service personnel, educational programs emphasizing food service care for persons with such infections are needed.[59]

Disposable utensils, plateware, or trays are not required for hospitalized patients infected with HIV, although proper dishwashing and cleaning procedures are paramount.[30,56] Food service establishments need to maintain adequate quality assurance standards to ensure that the proper sanitation procedures are used in food storage, preparation, and serving as well as the cleaning of the food service facilities.

REFERENCES

1. Archer, D., and Young F., Contemporary issues: diseases with a food vector, *Clin. Micro. Rev.,* 1, 377-398, 1988.
2. Jackson, G., Public health and research perspectives on the microbial contamination of foods, *J. Anim. Sci.,* 68, 884-891, 1990.
3. Roberts, D., Sources of infection: food, *Lancet,* 336, 859-861, 1990.
4. Casemore, D., Foodborne protozoal infection, *Lancet,* 336, 1427-1432, 1990.
5. Appleton, H., Foodborne viruses, *Lancet,* 336, 1362-1364, 1990.
6. Raiten, D., Nutrition and HIV infection: A review and evaluation of the existent knowledge of the relationship between nutrition and HIV infection, *Nutr. Clin. Prac.,* Suppl., 6, 1s-94s, 1991.
7. Smith, P., Lane, H., Gill, V., Manishewitz, J., Quinnian, G., Fauci, A., and Masur, H., Intestinal infections in patients with the acquired immunodeficiency syndrome (AIDS), *Ann. Intern. Med.,* 108, 328-333, 1988.
8. Lane, H., Masur, H., Gelmann, E., Longo, D., Steis, R., Chused, T., Whaton, G., Edgar, L., and Fauci, A., Correlation between immunologic function and clinical subpopulations of patients with the acquired immunodeficiency syndrome, *Am. J. Med.,* 78, 417-422, 1985.
9. Rodgers, V., Fassett, R., and Kagnoff, M., Abnormalities in intestine mucosal T-cells in homosexual populations including those with the lymphadenopathy syndrome and acquired immunodeficiency syndrome, *Gastroenterology,* 90, 552-558, 1986.
10. Perlman, D., Ampel, N., Schifman, R., Cohn, D., Patton, C., Aguire, M., Wang, W., and Blaser, M., Persistent *Compylobacter jejuni* infections in patients with HIV, *Ann. Intern. Med.,* 108, 540-546, 1988.
11. Current, W., and Garcia, L., Cryptosporidiosis, *Clin. Lab. Med.,* 11, 873-897, 1991.
12. Archer, D., and Glinsman, W., Enteric infections and other cofactors in AIDS, *Immunol. Today,* 6, 292-295, 1985.
13. Casemore, D., Sands, R., and Curry, A., *Cryptosporidium* species a "new" human pathogen, *J. Clin. Pathol.,* 38, 1321-1336, 1985.
14. Food Safety and Inspection Service, USDA, Preventing foodborne illness. A guide to safe food handling, *Home Garden Bull.,* no. 247, 1990.
15. Jacobs, J., Gold, J., Murray, H., Roberts, R., and Armstrong, D., *Salmonella* infections in patients with the acquired immunodeficiency syndrome, *Ann. Intern. Med.,* 102, 186-188, 1985.
16. Glaser, J., Morton-Kute, L., Berger, S., Weber, J., Siegal, F., Lopez, C., Robbins, W., and Landesman, S., Recurrent *Salmonella typhimurium* bacteremia associated with the acquired immunodeficiency syndrome, *Ann. Intern. Med.,* 102, 189-193, 1985.
17. Nadelman, R., Mathur-Wagh, V., Yancouitz, S., and Mildvan, D., *Salmonella* bacteremia associated with the acquired immunodeficiency syndrome (AIDS), *Arch. Intern. Med.,* 145, 1968-1971, 1985.
18. Fisch, M., Dickinson, G., Sinave, C., Pitchenik, A., and Cleary, T., *Salmonella* bacteremia as manifestation of acquired immunodeficiency syndrome, *Arch. Intern. Med.,* 146, 113-115, 1986.

19. Smith, P., Macher, A., Bookman, M., Boccia, R., Steis, R., Gill, V., Manischewitz, J., and Gelman, E., *Salmonella typhimurium* enteritis and bacteremia in the acquired immunodeficiency syndrome, *Ann. Intern. Med.,* 102, 207-209, 1985.

20. Profeta, S., Forrester, C., Eng, R., Liu, R., Johnson, E., Palinkas, R., and Smith, S., *Salmonella* infections in patients with acquired immunodeficiency syndrome, *Arch. Intern. Med.,* 145, 670-672, 1985.

21. Celum, C., Chaisson, R., Rutherford, G., and Berhart, J., Incidence of salmonellosis in patients with AIDS, *J. Infect. Dis.,* 156, 998-1002, 1987.

22. Mascola, L., Lieb, L., Chiu, J., Fannin, S., and Linnan, M., Listerosis: An uncommon opportunistic infection in patients with acquired immunodeficiency, *Am. J. Med.,* 84, 162-164, 1988.

23. Sherman, C., Raucher, B., Epstein, J., and Berger, M., Outpatient nutritional care, in *Quality Food and Nutrition Services for AIDS Patient,* ASPEN Publication, Rockville, MD, 129-138, 1990.

24. Archer, D., Food counseling for persons infected with HIV: Strategy for defensive living, *Pub. Health. Rep.,* 104, 196-198, 1989.

25. Griffin, P., and Tauxe, R., Food counseling for patients with AIDS, *J. Infect. Dis.,* 158, 668, 1988.

26. Filice, G., and Pomeroy, C., Preventing secondary infections among HIV-positive persons, *Pub. Health Rep.,* 106, 503-517, 1991.

27. Center for Disease Control, Food and Drug Administration: Eating defensively, food safety advice for persons with AIDS, Sept. 1989, Distributed by National AIDS Information Clearinghouse, telephone: 1-800-458-5231.

28. Dhundale, K., and Hubbard, P., Home care for the AIDS patient: Safety first, *Nursing,* 34, 34-36, 1986.

29. Schietinger, H., A home care plan for AIDS, *Am. J. Nutr.,* 86, 1021-1027, 1986.

30. Conte, J., Infection with human immunodeficiency virus in the hospital, *Ann. Intern. Med.,* 105, 730-736, 1986.

31. Berry, R., Home care of the child with AIDS, *Ped. Nurs.,* 14, 341-344, 1988.

32. Newman, C., Practical dietary recommendations, in *HIV Infection in Gastrointestinal and Nutritional Manifestations of the Acquired Immunodeficiency Syndrome,* Raven Press, New York, 269-270, 1991.

33. Farley, D., Food safety crucial for people with lowered immunity, *FDA Consumer,* July-Aug., 7-9, 1990.

34. Aker, S., and Chevey, C., The use of sterile and low microbial diets in ultraisolation environments, *JPEN,* 7, 390-397, 1983.

35. Ghiron, L., Dwyer, J., and Stollman, L., Nutrition therapy for AIDS: New directions, *Clin. Nutr.,* 8, 114-119, 1989.

36. De-Leeuw, I., and Vandewoude, M., Bacterial contamination of enteral diets, *Gut.,* 27 (suppl. 1), 56-57, 1986.

37. Schreiner, R., Eitzen, N., Gfell, M., Kress, S., Gresham, E., French, M., and Moye, L., Environmental contamination of continuous drip feeding, *Pediatrics,* 63, 232-237, 1979.

38. Allwood, M., Microbial contamination of parenteral and enteral nutrition, *Acta. Chir. Scand.,* 507, 383-387, 1979.

39. DeVries, E., Mulder, N., Houwen, B., and DeVries-Hospers, H., Enteral nutrition by nasogastric tube in adult patients treated with intensive chemotherapy for leukemia, *Am. J. Clin. Nutr.,* 35, 1490-1496, 1982.

40. Anderson, K., Norris, D., Godfrey, L., Avent, C., and Butterworth, C., Bacterial contamination of tube-feeding formulas, *JPEN,* 8, 673-678, 1984.
41. Freedland, C., Roller, R., Wolfe, B., and Flynn, N., Microbial contamination of continuous drip feedings, *JPEN,* 13, 18-22, 1989.
42. Paauw, J., Fagerman, K., McCarmish, M., and Dean, R., Enteral nutrition solutions: Limiting bacterial growth, *Am. Surg.,* 50, 312-316, 1984.
43. Fagerman, K., Paauw, J., McCamish, M., and Dean, R., Effects of time, temperature, and preservation on bacterial growth in enteral nutrient solutions, *Am. J. Hosp. Pharm.,* 41, 1122-1126, 1984.
44. Skipper, A., Monitoring and complications of enteral feeding, in *Dietitian's Handbook of Enteral and Parenteral Nutrition,* ASPEN Publishers, Rockville, MD, 293-309, 1989.
45. Silk, D., and Payne-James, J., Complications of enteral nutrition, in *Clinical Nutrition Enteral and Tube Feeding,* 2nd ed., WB Saunders Co., Philadelphia, PA, 510-531, 1990.
46. Gaile, M., Enteral feeding and infection in the immunocompromised patient, *Nutr. Clin. Prac.,* 6, 55-64, 1991.
47. Hoestetler, C., Lipman, T., Gerachty, M., and Parker, P., Bacterial safety of reconstituted continuous drip tube feeding, *JPEN,* 6, 232-235, 1982.
48. Bastow, M., Greaves, P., and Allison, S., Microbial contamination of enteral feeds, *Hum. Nutr. Appl. Nutr.,* 37A, 426-440, 1982.
49. Hill, D., and Pearson, R., Health advice for international travel, *Ann. Intern. Med.,* 108, 839-852, 1988.
50. Lange, W., and Denny, S., Travel in eastern Europe: Guidelines for patients, *Postgrad. Med.,* 89, 143-147, 1991.
51. Kozicki, M., Steffen, R., and Schar, M., Boil it, cook it, peel it or forget it: Does this rule prevent travelers' diarrhea?, *Int. J. Epidemiol.,* 14, 169-172, 1983.
52. Dickens, D., Dupont, H., and Johnson, P., Survival of bacterial enteropathogens in the ice of popular drinks, *J. Am. Med. Assoc.,* 253, 3141-3143, 1985.
53. Ericsson, C., Pickering, L., and Sullivan, P., The role of location of food consumption in the prevention of travelers' diarrhea in Mexico, *Gastroenterology,* 79, 812-816, 1980.
54. Center for Disease Control, Health information for international travel, Dept. of Health and Human Services, Public Health Services, HHS pub no. (CDC) 85-8280.
55. Henry, M., and Sneed, J., AIDS: A food service management concern, *School Food Serv. Res. Rev.,* 15, 6-11, 1991.
56. Center for Disease Control, Recommendation for preventing transmission of infection with HTLV-III/LAV in the workplace, MMWR, 34, 682-695, 1985.
57. Cross, E., AIDS: Legal implications for managers, *J. Am. Diet. Assoc.,* 92, 74-77, 1992.
58. Food Service Sanitation Manual, DNEW pub. no. (FDA) 78-2081, First printing June 1978, 1-96, 1976.
59. Collins, C., and Garcia, M., Nutrition intervention in the treatment of human immunodeficiency virus infection, *J. Am. Diet. Assoc.,* 89, 839-841, 1989.

Chapter

13

Specialized Nutrition Support

Cynthia Thomson

INTRODUCTION

The nutritional status of HIV-positive patients is frequently compromised. Weight loss, cachexia, protein deficiency and micronutrient deficiencies are commonly seen in both asymptomatic and symptomatic patients. The depression in nutritional status is thought to be multifactorial. Malnutrition in HIV disease may be associated with inadequate intake, hypermetabolism, altered metabolism, malabsorption or a combination thereof.[1-3] There is clear evidence that malnutrition will occur at some point in the disease process for greater than 95% of patients.[4] Approximately 65% will experience malabsorption; another 95% significant weight loss; and an additional 90% will have oral or esophageal infections which affect food intake levels. Anorexia is another common concern affecting intake. Anorexia may be related to psychological issues associated with the AIDS diagnosis, social isolation, biochemical changes including increased cytokine activity,[5] multiple medications, inactivity or opportunistic infections/fever.

Nutritional status has also been associated with clinical outcome.[6-8] McCorkindale et al.[9] reported a significant reduction in the life span of HIV-positive patients who had serum albumin levels below 2.5 mg/dl. A serum albumin level of greater than 3.0 has been associated with prolonged life span and decreased morbidity. Although data to date is limited in this area, retrospective data support the role of improved nutritional status and prolonged life with reduced morbidity. In addition, quality of life issues must be considered when formulating a medical/nutritional care plan. Many HIV-positive patients have demonstrated improved quality of life (increased activity, increased ability to perform activities of daily living, prolonged employment, etc.)[10] after the provision of adequate nutritional support.

0-8493-7842-7/94/$0.00+$.50
© 1994 by CRC Press, Inc.

The treatment of malnutrition associated with HIV disease requires not only a multidisciplinary approach, but also on-going nutritional status assessment and monitoring. Early intervention has been advocated to reduce the long-term effects of inadequate nutrition and the difficulty associated with replenishing tissue stores in patients prone to further nutritional losses. Initially, nutrition support should focus on nutrition education and use of the oral route to meet nutritional needs.[11,12] However, if this approach is unsuccessful, a more aggressive approach should be taken.[13-15] This chapter discusses the indications for and use of specialized nutritional support, both enteral and parenteral, in HIV-positive patients, as well as reviewing current standards of care for the provision and monitoring of specialized nutrition support in this patient population.

NUTRITIONAL STATUS ASSESSMENT

Components

The cornerstone of nutrition support is the nutritional status assessment. Nutritional status assessment is composed of four key areas of data collection: (1) anthropometric, (2) biochemical, (3) nutrition physical examination and (4) diet history. Serial anthropometric measurements, taken over several months duration, will allow clinicians to make the most accurate assessment. Anthropometric data are of critical importance to the assessment of nutritional status in HIV-positive individuals. Many HIV-positive patients will present with usual body weights below standard recommended levels, and therefore prompt attention to weight status is indicated. Kotler et al. has published several research reports suggesting appropriate parameters to be used in assessing the body composition of the HIV-positive patient.[16-18] Recently Kotler determined that many traditional methods for assessing body fat measurement in HIV patients are likely inaccurate. In his study of 18 male patients with AIDS he found total body water (a method widely available) and a newer method, dual-photon absorptiometry (DPA),[19] to be the most reliable methods for determining body composition. Anthropometric data may also be altered in the presence of dehydration or fluid retention, common concerns in AIDS patients.

Biochemical evaluation of nutritional status should, at minimum, include routine monitoring of serum albumin, transferrin, total protein, serum cholesterol and hemoglobin/hematocrit. Prealbumin levels will be useful in determining the short-term effectiveness of nutritional intervention. Depletion in any of the above parameters is indicative of either protein deficiency or anemia. Standard nutritional status indicators such as total lymphocyte count and skin test antigens are not useful in patients who are HIV positive due to the disease's impact upon immune markers.

The nutrition physical examination is used to evaluate the patient for signs of nutritional deficiencies. In particular, B vitamins, vitamin C, vitamin K, iron, energy and protein deficiencies can be manifested in the patient's physical appearance in the form of nutrient-based lesions.[20]

The diet history is of critical importance to the nutritional assessment process. Dietary intake will generally direct the nutrition physical evaluation, as well as provide support for anthropometric and biochemical findings. An initial diet history should be followed by routine re-evaluation of intake either through food intake records or regular re-assessments of 24-hour intake. Evaluation of medication regimes, including alternative therapies, will allow for a more comprehensive assessment, since many drugs will affect nutrient intake, absorption and assimilation.

Micronutrient Deficiencies

Micronutrient deficiencies are frequently seen in even the asymptomatic HIV-positive patient despite what appears to be adequate dietary intake. Deficiencies in vitamin B_{12}, B_6, A, E, and copper have been frequently cited.[21-24] Folate levels have been shown to be normal, elevated, and depleted indicating the wide variability which can exist among HIV patients. Thiamin, riboflavin, niacin and vitamin C levels have been shown to be adequate in HIV-positive patients.[25,26] Selenium and zinc deficiencies have also been established within the HIV population.[27,28] Routine vitamin/mineral supplementation for HIV-positive patients has not gained universal acceptance. Increasing scientific evidence of the prevalence of deficiencies in HIV patients has resulted in more clinicians advising patients to supplement their diets appropriately. For further information on micronutrient status in HIV, the reader is referred to a more comprehensive review by Coodley and colleagues.[21]

Establishing a Nutritional Care Plan

Once the nutritional assessment data have been collected, the nutritional care plan can be established. The assessment will address not only the adequacy of the diet, current nutritional status and need for change, but also further tests or procedures which would be indicated to assure the most reliable assessment possible. The goals for nutrition support in HIV-positive patients include: (1) minimize catabolism/control weight loss, (2) minimize and/or replenish nutrient losses, (3) replenish visceral and somatic protein mass, (4) maximize immune response/improve host resistance to opportunistic infections, (5) improve response to drug therapy, (6) enhance quality of life, and in doing so, cause no harm to the patient. Each nutritional care plan should be individualized to meet the specific needs of the patient. It is imperative that the HIV-positive individual contribute to the development of the nutritional care plan in an effort to optimize patient commitment to and compliance with the specific strategies prescribed.

Team Approach

The past two decades have brought attention to the efficacy of a medical team approach to patient care. Treating patients with HIV disease is no exception. When the overall patient care plan is multidisciplinary in development and implementation, compliance and improved clinical outcome are the more likely result. Team members might include any or all of the following: medical doctor, nurse, clinical dietitian, physical therapist, case worker, patient advocate, psychologist, social worker and clinical pharmacist. Use of the multidisciplinary team to evaluate, prescribe and monitor enteral and parenteral nutrition support has become the standard of care in the majority of hospitals.

Nutritional Requirements

Assessing the nutritional requirements of HIV-positive and AIDS patients is difficult. Current data on metabolic needs of this patient population have been conflicting. Investigators have reported decreased metabolic rates in asymptomatic patients.[29] Studies of AIDS patients have shown increases in metabolic rate from 10 to 30% above control patients.[30] Clearly, the presence of infection, fever or malignancies will result in increased energy requirements. The wasting syndrome associated with AIDS is thought to be partially related to hypermetabolism — but the extent to which metabolic rate plays a role appears to be patient specific. Clinicians currently employ the Harris Benedict formula for calculating basal energy expenditure and add to it a percentage increase for activity and metabolic stress. Until further clinical studies are completed in the HIV-positive population, the Harris Benedict is probably the most reliable "guesstimate" of energy consumption in HIV-positive patients.

Protein requirements are also not clearly established. Generally it is suggested that patients with HIV disease receive higher protein diets than what the RDA would suggest. Asymptomatic patients should be prescribed between 1.0 and 1.25 g/kg/d. Patients with increased metabolic stress should be prescribed at least 1.5 g/kg/d. Many metabolically stressed patients will require upwards of 2.0 to 2.5 g/kg/d to maintain positive nitrogen balance. During severe metabolic stress, positive nitrogen balance is generally not possible and the goal should be to minimize net losses in protein stores.

ENTERAL NUTRITION SUPPORT

Enteral nutrition refers to the provision of nutrients to the body via the gastrointestinal tract. In clinical practice this is assumed to include feedings via specialized feeding tubes such as the gastrostomy, jejunostomy, nasogastric, and nasoduodenal. Use of the gastrointestinal tract to deliver nutrients is not only more physiological than parenteral feeding, but appears to enhance nutrient absorption and utilization in a more cost effective manner.[31]

Indications

The indications for the use of enteral feeding of patients with AIDS are similar to non-AIDS patients and include: (1) significant malnutrition with functional gastrointestinal tract, (2) partial or upper gastrointestinal obstruction (lymphoma of the gastrointestinal tract), (3) severe dysphagia or esophagitis, (4) severe dementia, (5) coma, (6) documented malnutrition with inadequate volitional intake, (7) prolonged intake level of less than 50% of estimated needs, and (8) distal, low output enterocutaneous fistulas. Patients undergoing radiation and chemotherapy or patients with pancreatitis have also been successfully nourished using enteral nutrition. Enteral nutrition should not be the feeding route selected when HIV-positive patients are diagnosed with intestinal obstruction, ileus, severe/uncontrolled diarrhea, shock, high output enterocutaneous fistula or when patient prognosis or wishes do not support its implementation.[31] Recent research has shown that enteral nutrition is beneficial over parenteral nutrition in that enteral nutrition appears to preserve gut function and maintain gut integrity, thus reducing the risk of translocation of gut bacteria into the lymphatic system.[32]

Formula Selection

The decision as to which formula to use in enterally feeding HIV-positive patients can be a complex one. Currently there are well over one hundred nutritional formulas available. Generally formulas can be divided into three categories — intact protein/polymeric, elemental/peptide-based and specialized.

Intact/polymeric formulas have sodium or calcium caseinate as the protein source, medium and long chain triglycerides as a fat source, and sucrose, corn syrup or maltodextrin as the carbohydrate source. These formulas generally provide 100% of the RDAs in 1500 to 2000 mls. Polymeric formulas should be prescribed to patients with normal gut function.

Elemental or peptide based formulas have been shown to be efficacious in nutritionally repleting patients with bowel disease or malabsorption. This would include HIV-positive patients with pancreatitis, diarrhea, low output enterocutaneous fistulas, AIDS enteropathy and poor tolerance to polymeric formulas.[33]

The specialized formulas vary considerably in their application. Some have been designed for patients with renal or hepatic failure (although research has provided controversial reports on their efficacy); others are designed for use in patients with compromised pulmonary function; others for patients requiring increased nitrogen related to critical illness or metabolic stress.

Recently, pharmaceutical companies have begun to develop immune-stimulating enteral formulations. These enteral products contain supplemental levels of nutrients such as arginine, glutamine, omega three fatty acids, antioxidants, taurine and nucleotides — all of which have been advocated for their immune-enhancing properties.[34-36] The clinical data to support the use of specialized nutritional formulations remains controversial. Recent advances in the use of immunomodulating formulations appear promising, but published reports on their effectiveness in the HIV-positive population are not available. When selecting the most appropriate formula, nutritional requirements, tolerance issues, bowel status, feeding route, product availability and cost are all considerations. In many circumstances the formula selected to initiate the feeding will be not be the same as the final formula prescribed since the patient's clinical condition will likely change with time.

HIV-positive individuals will generally require an isotonic, low fat, lactose-free formula for optimal tolerance. High nitrogen content formulas are commonly prescribed to meet increased protein requirements.

Selection of a Feeding Route

HIV patients may be fed enterally through a variety of administration routes. The most commonly employed feeding route is the nasogastric route. Nasogastric tubes can be easily placed without anesthesia, and feeding via this route is generally well tolerated in the alert patient with normal gastric emptying. The second route of choice is the nasoduodenal route. Nasoduodenal feedings are used in the less alert patient or other patients at risk for aspiration. Surgically placed feeding tubes including gastrostomies and jejunostomies are used in patients who require long-term nutrition support. There is increased risk to the patient with placement of these tubes due to the requirement for anesthesia. Some HIV-positive patients will prefer a surgically placed feeding tube as it is less visible and allows for a more normal lifestyle. Percutaneous endoscopic gastrostomies (PEG) are often used instead of standard gastrostomies since they require only local anesthesia for placement and feeding can be initiated within 24 hours. Surgically placed tubes can, however, be a locus for bacterial invasion.

Administration

Bolus or continuous feedings are prescribed for use in patients with gastric feeding tubes while patients with duodenal or jejunal feedings should be fed using a continuous feeding regime for best tolerance. Bolus feedings are thought to be more physiologic, less time intensive to feed, and allow the patient more flexibility in their daily schedule. Continuous feedings are generally better tolerated particularly by the compromised gastrointestinal tract or in patients who have had a prolonged period of gut rest. Patients being fed into the lower bowel can adjust to bolus feedings, but this will require a gradual transitional feeding plan and close monitoring for tolerance. Nocturnal feedings are advised in patients who require enteral feeding to augment inadequate oral intake.

Enteral formulas which are isotonic (300 mOsm/kg H_2O) can be initiated at full strength. Starting with full strength formula will allow for more expedient progression to full nutritional support. Formulas which are hypertonic should be diluted with water to half or quarter strength for initial administration. Once tolerance is demonstrated the rate *or* strength can be increased accordingly. Increasing both the rate and strength simultaneously is not advised as tolerance is usually compromised as a result.

Closed enteral feeding delivery systems are advised for use in the HIV-positive patient population whenever possible. Closed systems have been shown to reduce risk of contamination.[37]

Monitoring of Enteral Feeding

As with any medical therapy, the patient receiving enteral support will require close monitoring for tolerance, biochemical and fluid balance, and maintenance or improvement in nutritional status. Tolerance should be evaluated using parameters such as: consistency and frequency of stooling, abdominal distention, reflux/aspiration and gastric residuals.

Biochemical monitoring should include weekly albumin and transferrin, and biweekly prealbumin to assess adequacy of protein support. Serum sodium, potassium, chloride, bicarbonate, glucose, blood urea nitrogen, creatinine, calcium, magnesium and phosphorus should be evaluated daily until stable then two to three times weekly. Patients receiving home enteral support should have a complete biochemical evaluation at least monthly and more frequently if their clinical condition is less stable. HIV-positive patients who present with significant weight loss and malnutrition are at risk for "refeeding syndrome",[38-39] a clinical syndrome in which electrolyte abnormalities occur in relation to increased cellular requirements related to glucose administration. If not monitored closely with aggressive electrolyte replacement (as indicated by laboratory values), refeeding syndrome can be life-threatening.

Efficacy of Enteral Support in HIV Patients

The use of enteral nutrition to reverse malnutrition has gained widespread acceptance in the clinical setting. To date, however, studies addressing the efficacy of enteral support in HIV-positive patients remain scant. Ferraro et al.[40] were able to demonstrate improved nutritional status in five AIDS patients given two months of enteral nutrition support. Kotler found similar results in eight severely ill, hospitalized HIV-positive patients. His study also suggested that enteral support is associated with improved mental status and ability to discharge the patient.[41] Low fat, elemental formulas have been prescribed in HIV-positive patients with documented diarrhea and malabsorptive syndromes with resultant improvement in weight status as well as a reduction in bowel movements.[42] One could postulate that reduction in bowel movements would in turn result in fluid balance and improved electrolyte retention. More research is needed to establish specific standards of care for the use of enteral support in HIV-positive patients.

PARENTERAL NUTRITION

Parenteral nutrition refers to intravenous administration of nutrients. Generally infusion occurs through the subclavian, jugular or peripheral veins.

Indications

Parenteral nutrition is indicated for HIV-positive patients who demonstrate intolerance to enteral feedings, have been diagnosed with bowel obstruction, severe pancreatitis, copious/high volume diarrhea, intractable vomiting, radiation enteritis, high output enterocutaneous fistula or in malnourished patients who will require prolonged bowel rest. Parenteral nutrition should not be used in HIV-positive patients who have functional and usable gastrointestinal tracts and are capable of adequately absorbing nutrients enterally. Total parenteral nutrition should also not be used unless the patient is in agreement, and the prognosis warrants aggressive nutritional intervention.[43] Certain patients with end stage AIDS might be better supported with intravenous 5% dextrose, vitamins/minerals or Procalamine (McGaw Pharmaceutical) — a dextrose solution with minimal electrolyte supplementation, to meet hydration and basic physiological needs without aggressive nutrition support.

Peripheral versus Central Parenteral Nutrition

Patients who require only short-term parenteral support, less than seven days, can be fed via the peripheral route. However, some AIDS patients may have poor peripheral access due to previous chemotherapy, intravenous drug use, or repeated blood transfusions. The peripheral route will limit tolerance for high osmolality and therefore the concentration of the solution administered. Hypertonic solutions of greater than 800 mOsm have been associated with increased risk of phlebitis if administered peripherally. If prolonged parenteral support is indicated, a central line is indicated. In patients requiring home parenteral support a more permanent indwelling central access catheter should be surgically placed. Patients' caloric, protein and micronutrient needs can be met peripherally, but a larger volume of daily infusate will be required. All patients, but particularly HIV-positive patients, should have a new central or peripheral line placed when initiating parenteral support. This practice will reduce the risk of infection.

Selection of a Feeding Solution

Determination of the patient's energy, protein, fluid and micronutrient requirements will direct the final parenteral solution prescribed. Some hospitals will allow for flexibility in the solution components, while others will have established standard solutions. Occasionally, HIV-positive individuals will present with hypertriglyceridemia, making the decision as to whether to provide lipid calories intravenously a difficult one. Clinically, few HIV-positive patients with elevated triglyceride levels will experience a continued rise in serum levels with intravenous lipid administration, unless the energy from fat accounts for greater than 60% of total calories. Therefore, in most patients, elevated serum triglyceride levels should not preclude the use of intravenous lipid. Glucose intolerance may occur in HIV-positive patients who have significant metabolic stress. If glucose intolerance is seen, regular insulin can be added to the parenteral nutrition solution directly or dextrose calories can be replaced by lipid. Adequate micronutrient provision is essential, particularly in patients at risk for refeeding syndrome. Supplemental zinc, copper, folate, B_{12} and B_6 are advisable above standard vitamin/trace element administration levels.

Administration

Peripheral parenteral nutrition, with its lower dextrose concentration, can generally be initiated at 80 to 110 ml/h without resultant hyperglycemia. Central parenteral nutrition with dextrose concentrations from 15 to 35% should be initiated at 30 to 40 ml/h (not to exceed 60 ml/h) and the infusion rate increased every 12 to 24 hours as indicated by laboratory results. Parenteral nutrition should always be administered on a continuous basis (not bolus) to avoid electrolyte imbalances. Cyclic infusions can be established over time and with careful monitoring. When weaning patients off parenteral nutrition, solutions with dextrose concentrations higher than 10% require a step-wise decrease in administration rate over 3 to 6 hours duration to avoid hypoglycemia.

Parenteral nutrition access lines should be reserved for the administration of parenteral nutrition so that contamination can be prevented. Infection rates correlate with the total number of line manipulations; therefore, line contact should be kept to a minimum.

Monitoring

Parenteral nutrition requires close monitoring to assure patient safety and to avoid complications which have been associated with its use. Nursing and medical staff are responsible for evaluating the proper placement and patency of either peripheral or central venous catheters. Problems such as pneumothorax, hemothorax and brachial nerve palsy are associated with line placement. Air embolism can occur at any time while the catheter is in place. Extravascular malpositioning of the catheter has also occurred as has leaking, catheter damage, clogging or infiltration. Each of these complications can be diagnosed with a routine X-ray or upon clinical evaluation.

Biochemical monitoring should include at least daily serum sodium, potassium, chloride, bicarbonate, glucose, blood urea nitrogen, creatinine, calcium magnesium, phosphorus, hemoglobin, hematocrit, and white count with differential until the patient is stable at goal infusion rate. Nutritional status should be evaluated using weekly albumin and transferrin, biweekly prealbumin levels. Patients on home parenteral nutrition support should have at least biweekly evaluation of the above laboratory values, more frequently if the clinical status dictates.

Daily weights, intake and output and (every shift) vital signs will allow for a complete clinical assessment. Line access sites should be evaluated routinely and dressing changes should occur in a timely manner and according to hospital protocol.

Risk versus Benefit

One of the key issues in the provision of parenteral nutrition to patients with AIDS is the issue of infection. Total parenteral nutrition can lead to systemic infections in non-HIV-positive patients; therefore, there exists concern that placement of a central line and infusion of intravenous nutrients may present a heightened risk in this patient population. Singer et al. reviewed the incidence of infection in a group of AIDS patients on home, long-term parenteral support and found no increase in infection rates in AIDS versus non-AIDS patients.[44] However, sparse data is available at this time.

The efficacy of parenteral nutrition support in HIV-positive patients is also under investigation. Kotler found supplementation with parenteral nutrition had variable effects on body composition.[45] Repletion occurred in patients with eating disorders or malabsorption, but continued nutritional depletion occurred in patients with serious systemic infections. A study by Brolin and colleagues at Robert Wood Johnson University Hospital showed no measurable benefit to the use of supplemental nutrition support, either parenteral or enteral. The study also demonstrated that aggressive nutrition support (enteral and parenteral) was only provided to patients with an extremely poor clinical status. This study also suggested that patients given enteral support had an improved outcome when compared with patients receiving parenteral support.[46] Other studies have shown parenteral nutrition in HIV-positive patients to be associated with increased fat stores in the majority of patients, but increased protein mass in only a few.[47] Further studies of the efficacy of nutrition support in this patient population are needed.

SUMMARY

Malnutrition is apparent in the majority of AIDS patients as well as a significant number of HIV-positive individuals. Given ample nutritional counseling, most patients can achieve desired nutritional goals. However, a percentage of the HIV-positive population will require more aggressive nutritional intervention in the form of enteral or parenteral support to meet their nutritional requirements. Currently there is no optimal formulation for all HIV-positive patients. Although the development of immune-stimulating products may hold promise for the future, further clinical trials specifically involving HIV-positive patients are needed before their use can be recommended. Both enteral and parenteral support can be administered safely and have resulted in weight gain in HIV-positive individuals. However, only enteral nutrition has resulted in appreciable improvement in visceral protein status. Clearly patients with end stage disease are not likely to demonstrate a change in clinical course related specifically to the provision of nutritional support. Early intervention in patients with significant weight loss and inability to meet nutritional requirements by mouth is critical to achieving optimal nutritional care goals. Development of rational strategies for nutrition support of patients with AIDS should be a high priority.[48]

REFERENCES

1. Hellerstein, M.K., Kahn, J., Mudie, H. and Viteri, F., Current approach to the treatment of human immunodeficiency virus-associated weight loss: Pathophysiologic considerations and emerging management strategies, *Seminars in Oncol,* 17(suppl 9), 17, 1990.
2. Hecker, L.M. and Kotler, D.P., Malnutrition in patients with AIDS, *Nutr Rev,* 48, 1990.
3. O'Sullivan, P., Linke, R.A. and Dalton, S., Evaluation of body weight and nutritional status among AIDS patients, *J Am Diet Assoc,* 85, 1483, 1985.
4. Chelluri, L. and Jastremski, M.S., Incidence of malnutrition in patients with acquired immunodeficiency syndrome, *Nutr Clin Pract,* 4, 16, 1989.
5. Grunfeld, C. and Feingold, K.R., Metabolic disturbances and wasting in the acquired immunodeficiency syndrome, *Semin in Med of Beth Israel Hospital,* 327, 329, 1992.
6. Ysseldyke, L.L., Nutritional complications and incidence of malnutrition among AIDS patients, *J Am Diet Assoc,* 90 (suppl), A-55, 1990.
7. Falutz, J., Tsoukas, C. and Deutsch, G., Functional correlates of decreased serum zinc in human immunodeficiency virus disease (abstract), Proceedings of the 5th International Conference on AIDS, June 4-9, Montreal, 468, 1989.
8. Kotler, D.P., Tierney, A.R., Wang, J. and Pierson, R.N., The magnitude of tissue wasting and timing of death in patients with AIDS, *Clin Res,* 35, 368, 1987.
9. McCorkindale, C., Dybevik, K., Coulston, A.M. and Sucher, K.P., Nutritional status of HIV-infected patients during the early disease stages, *J Am Diet Assoc,* 90, 1236, 1990.
10. Broderick, B. and Nesset, J., Nutritional status parameters associated with psychosocial well being seen in AIDS/ARC clients at an outpatient clinic in San Francisco, Proceedings of the 5th International Conference on AIDS, June 4-9, Montreal, 469, 1989.
11. Smerko, A., Oral nutritional products for the HIV outpatient, *AIDS Patient Care,* 4, 17, 1990.
12. Position of the American Dietetic Association: Nutrition intervention in the treatment of human immunodeficiency virus infection, *JADA,* 839, 1989.

13. McQuiggan, M., Enteral nutrition for the hospitalized HIV patient, *AIDS Patient Care,* 4, 13, 1990.

14. Kordura, M.J., Guenter, P. and Rombeau, J.L., Enteral nutrition in the critically ill, *Crit Care Clin,* 3, 133, 1987.

15. Thomson, C.T., Specialized nutrition support in: Issues and Interventions/Nutrition and HIV Disease, The Cutting Edge, Denver, Colorado, October 14, 1989.

16. Kotler, D.P., Wang, J. and Pierson, R.N., Body composition studies in patients with acquired immunodeficiency syndrome, *Am J Clin Nutr,* 42, 1255, 1985.

17. Mantero-Atienza, E., Beach, R.S., Fletcher, M., Van Riel, F., Morgan, R., Eisdorfer, C. and Fordyce-Baum, M.K., Measures of nutritional status in early HIV infection, *Arch AIDS Res,* 3, 275, 1989a.

18. Garcia, M., Collins, C.L. and Mansell, P.A., The acquired immunodeficiency syndrome: Nutritional complications and assessment of body weight status. *Nutr Clin Pract,* 2, 108, 1987.

19. Wang, J., Kotler, D.P., Russell, M., Burastero, S., Mazariegos, M., Thornton, J., Dilmanian, F.A. and Pierson, R.N., Body-fat measurement in patients with acquired immunodeficiency syndrome: which method should be used?, *Am J Clin Nutr,* 56, 963, 1992.

20. Kight, M.A., The nutrition physical examination, *CRN Quart,* 2, 1, 1987.

21. Coodley, G. and Girard, D.E., Vitamins and minerals in HIV infection, *J Gen Int Med,* 6, 472, 1991.

22. Beach, R.S., Mantero-Atienza, E. and Fordyce-Baum, M.K., Dietary supplementation in HIV infection (abstract), *FASEB J,* 2, A1435, 1989.

23. Harriman, G.R., Smith, P.D., Horne, M.K., Fox, C.H., Koenig, S., Lack, E.E., Lane, H.C. and Fauci, A.S., Vitamin B_{12} malabsorption in patients with acquired immunodeficiency syndrome, *Arch Intern Med,* 149, 2039, 1989.

24. Baum, M.K., Proceedings of The New York Academy of Sciences, Feb 10-12, Washington, D.C., 1992.

25. Mantero-Atienza, E., Beach, R.S., Van Riel, F., Morgan, R., Eisdorfer, C. and Fordyce-Baum, M.K., Low vitamin B_6 levels and immune dysregulation in HIV-1 infection (abstract), Proceedings of the Fifth International Conference on AIDS, June 4–9, Montreal, 468, 1989.

26. Bogden, J.D., Frank, O., Perez, G., Kempf, F., Bruening, K. and Louria, D., Micronutrient status and human immunodeficiency virus (HIV) infection, *Ann NY Acad Sci,* 587, 189, 1990.

27. Galvin, T.A., Micronutrients: Implications in human immunodeficiency virus disease, *Top Clin Nutr,* 7, 63, 1992.

28. Dworkin, B.M., Rosenthal, W.S., Wormser, G.P. and Weiss, L., Selenium deficiency in the acquired immunodeficiency syndrome, *J Par Ent Nutr,* 10, 405, 1986.

29. Kotler, D.P., Tierney, A.R., Brenner, S.K., Couture, S., Wang, J. and Pierson, R.N., Preservation of short-term energy balance in clinically stable patients with AIDS, *Am J Clin Nutr,* 51, 7, 1990.

30. Melchoir, J.C., Salmon, D., Rigaud, D., Leport, C., Bouvet, E., Betruchis, P., Vilde, J.L., Vachon, F., Coulaud, J.P. and Apfelbaum, M., Resting energy expenditure is increased in stable, malnourished HIV-infected patients, *Am J Clin Nutr,* 53, 437, 1991.

31. American Society of Parenteral and Enteral Nutrition Board of Directors, Guidelines for the use of enteral nutrition in the adult patient, *J Parent Ent Nutr,* 11, 435, 1987.

32. Andrassy, R.J., Preserving gut mucosal barrier and enhancing immune response, *Contemp Surg,* 32, 1, 1988.

33. Hickey, M.S., *Handbook of Enteral, Parenteral and ARC/AIDS Nutritional Therapy,* Mosby Year Book, St. Louis, 1992.

34. Daly, J.M., Lieberman, M.D., Goldfine, J., Shou, J., Weintraub, F., Rosato, E.F. and Lavin, P., Enteral nutrition with supplemental arginine, RNA, and omega-3 fatty acids in patients after operation: Immunologic, metabolic and clinical outcome, *Surgery,* 112, 56, 1992.

35. Cerra, F.B., Lehman, S., Konstantinides, N., Konstantinides, F., Shronts, E.P. and Holman, R., Effect of enteral nutrient on *in vitro* tests of immune function in ICU patients: a preliminary report, *Nutrition,* 6, 84, 1990.

36. Barbul, A., Arginine: biochemistry, physiology and therapeutic implications, *J Parent Ent Nutr,* 10, 227, 1986.

37. Vaughan, L.A., Manore, M. and Winstow, D., Bacterial safety of closed-administration system for enteral nutrition solutions, *J Am Diet Assoc,* 88, 35, 1988.

38. Apovian, C.M., McMahon, M. and Bristrian, B.R., Guidelines for refeeding the marasmic patient, *Crit Care Med,* 18, 1030, 1990.

39. Havala, T. and Shronts, E., Managing the complications associated with refeeding, *Nutr Clin Pract,* 5, 23, 1990.

40. Ferraro, R., Kotler, D.P., Cuff, P., Tierney, A.R., Smith, R. and Heymsfield, S., Effect of enteral nutritional therapy on body cell mass in AIDS (abstract), *Clin Res,* 37, 330A, 1989.

41. Kotler, D.P., Tierney, A.R., Ferraro, R., Cuff, P., Wang, J., Pierson, R.N. and Heymsfield, S., Enteral alimentation and repletion of body cell mass in malnourished patients with acquired immunodeficiency syndrome, *Am J Clin Nutr,* 53, 149, 1991.

42. King, A.B., McMillan, G., St. Arnaud, J. and Ward, T.T., Less diarrhea seen in HIV-positive (HIV +) patients on low-fat, elemental diet (abstract), Fifth International Conference on AIDS, June 4–9, Montreal, 1989.

43. Hecker, L.M. and Kotler, D.P., Malnutrition in patients with AIDS, *Nutr Rev,* 48, 393, 1990.

44. Singer, P., Rothkopf, M.M., Kvetan, V., Kirvela, O., Gaare, J. and Askanasi, J., Risks and benefits of home parenteral nutrition in the acquired immunodeficiency syndrome, *J. Parent Ent. Nutr.,* 15, 75, 1991.

45. Kotler, D.P., Tierney, A.R., Culpepper-Morgan, J.A., Wang, J. and Pierson, R.N., Effect of home total parenteral nutrition on body composition in patients with acquired immunodeficiency syndrome, *J Parent Ent Nutr,* 14, 454, 1990.

46. Brolin, R.E., Gorman, R.C., Milgrim, L.M., Abbott, J.M., George, S. and Gocke, D.J., Use of nutrition support in patients with AIDS, *Nutrition,* 7, 19, 1991.

47. Culpepper-Morgan, J.A., Kotler, D.P., Tierney, A.R., Wang, J. and Pierson, R.N., Efficacy of parenteral nutrition therapy of malnutrition in AIDS (abstract), *Clin Res,* 36, 355A, 1988.

48. Aron, J.M., Toward rational nutritional support of the human immunodeficiency virus-infected patient, *J. Parent Ent. Nutr.,* 15, 121, 1991.

Chapter

14

Optimization of Nutritional Support in HIV Disease

Jeffrey M. Aron

NATURAL HISTORY OF HIV INFECTION: A NUTRITIONAL POINT OF VIEW

HIV infection has been the most intensely studied viral infection in history.[1] A careful dissection of the virologic and immunologic pathophysiology has allowed for a clearer understanding of the clinical course of this disease.[1]

It has been long recognized that malnutrition is an inexorable co-morbidity of HIV infection,[2] and that significant loss of lean body mass which proceeds more rapidly than the clinical appearance of weight loss will lead to a patient's death regardless of the coexistence of any AIDS-defining illness.[3] However, there is still no convincing evidence that nutritional intervention, especially with such expensive modalities as total parenteral nutrition (TPN), has reduced mortality. Nonetheless, an overview of the virologic, immune, metabolic and clinical features as HIV disease evolves will provide a framework to optimize nutritional intervention. Even though our knowledge of the metabolic permutations in HIV disease is expanding, it remains rudimentary and our nutritional interventions even more so. We must therefore resort to logic and prudence, tempered by the few existing scientific observations to provide optimal nutritional support to our HIV patients.

0-8493-7842-7/94/$0.00+$.50
© 1994 by CRC Press, Inc.

EARLY HIV INFECTION

Viral and Immunologic Events

The HIV virus gains access to the host as free virus or bound within circulating lymphocytes or monocytes. The blood stream with or without mucosal sites is accessed with regional dissemination to lymph nodes or other lymphoid tissue, such as the gut-associated lymphoid tissue (GALT).[1] The volume of HIV and the mutated state of the virus from the donor will affect the subsequent course of the infection in the new host. It has been observed that those infected earlier in the epidemic have a longer latent period and a less virulent course — so-called "slow progressors" — than those infected later on who may proceed to AIDS more rapidly. This could be related to a sustained cytotoxic T cell response, more effective anti-HIV antibodies and a greater suppressor T cell response inhibiting HIV reproduction in CD4 cells.[4-7] Some of these responses could be governed by cytokine activity. Thus, the evolution of the disease may vary on an individual basis and affect the metabolic and nutritional parameters differently, rendering nutritional intervention a highly individualized process.

Weight change in HIV disease is not a continuing process, but rather is characterized by periods of loss and regain that vary depending upon the amount of lymphoid and macrophage tissue reservoirs harboring HIV.[1] Cytokine elaboration occurs with elevated levels of TNF alpha, TNF gamma, interleukin-1, interleukin-6, interferon alpha, beta, gamma[8-12] and various tissue growth factors (GCSF, MGCSF, PAF).[13,14]

Metabolic and Nutritional Events

In early HIV infection when the number of circulating CD4 cells are normal,[15] a 10% increase in resting energy expenditure (REE) has been demonstrated[15] along with a diminished fat free mass.[16,17] Triglycerides are slightly elevated[16] while levels of the counterregulatory hormones (catecholamines, glucagon, cortisol and growth hormone) are normal,[18] leading to the suspicion that the cytokine elevations are responsible for this redistribution of body mass compartments. Indeed, elevated levels of neopterin, a macrophage-derived protein resulting from interferon gamma stimulation,[12] early in HIV disease, has been found to predict wasting in HIV disease more accurately than CD4 levels. This strongly suggests that immune activation is a major determinant of weight loss.[19] While some observers cast doubt over the direct correlation of particular cytokines such as TNF alpha to wasting[20] it appears that the contextual totality of cytokine interaction is responsible for these changes. Thus, in early HIV infection, the setting for the wasting that continues throughout the disease is well established (Table 1).

TABLE 1
Metabolic and Nutritional Events
in Early HIV Disease

REE increased 8–10%
Decreased fat free (i.e., lean body) mass
Elevated triglycerides

Nutritional Interventions in Early HIV Disease

Guidelines have been established for nutritional support in this setting[21,22] which focus on improving oral intake with nutrient-dense food supplements along with vitamin supplementation (Table 2).

Certainly these guidelines are prudent and should be followed. Important at this point is the referral to a nutritional support dietician knowledgeable in HIV disease. Counseling on nutrient selection and food preparation should begin and be periodically reinforced. This will establish a strong link between nutrition and HIV infection in the mind of the patient and physician that will alert both parties to more aggressive nutritional intervention when needed later. An optimal strategy would include measurements of energy expenditure and body composition to further the awareness of nutritional changes (Table 3), along with a regular program of exercise which can favor restoration of lean body mass.[23]

As Hellerstein's group has shown, repletion of weight loss even in stable, asymptomatic HIV patients results in quantitatively greater gain of body water than lean body mass,[24] so while weight can be regained, the relative proportion of lean body mass lessens. This phenomenon is frequently hidden from the eye of the patient and practitioner, leading both to feel more secure that weight has been regained and "all is well again" for the time being.

Experimental Approaches to Nutritional Management in Early HIV Disease

Changing the total context of cytokine response early on in HIV infection should be a high priority for research. While antibodies to cytokines in acute bolus doses have reversed some of the abnormal biochemical parameters of wasting in animal cancer experiments,[25] and drugs such as pentoxifylline and thalidomide[27] inhibit individual cytokine production, the former appear to be impractical for widespread use in AIDS, and the latter need to be tested in controlled experiments. Studies are ongoing in cancer cachexia and advanced AIDS with these agents. They deserve to be studied *early* in the disease.

However, the seminal work of Endres'[28] group has demonstrated that cytokine context can be changed by altering eicosanoid metabolism via the

TABLE 2
Established Guidelines for Nutritional Support in
Early HIV Disease

Maximize Food Intake
• Substitute calorie-containing and nutrient-dense foods and
   beverages for low or no-calorie containing foods and
   beverages
• Increase number and size of meals daily
• Fortify foods with calorie and protein-containing ingredients
• Use calorie-containing condiments
• Modify diet according to tolerance
• Add calorie-containing supplements as needed
• Supplement vitamins — especially B6, beta-carotene, B_{12}
• Supplement trace elements — especially selenium

After Fields-Newman, C., *Nutrition and HIV/AIDS,* 313, 1289, 1992.

TABLE 3
Optimal Nutritional Interventions in
Early HIV Disease

• Involve a dietician conversant with HIV disease
• Measure or estimate energy expenditure and
   provide increased intake accordingly
• Measure body composition
• Involve patient in active exercise programs

feeding of w-3 fatty acids (fish oils). TNF alpha, IL-1 alpha and beta decreased after 6 weeks of fish oil feedings, 18 grams per day, and persisted for 10 weeks after their discontinuation.[28] However, Hellerstein did not show improvements in food intake, weight and lean body mass when 20 AIDS patients were given 18 grams of fish oil for 10 weeks, even though triglycerides fell.[29] Again, studies of these agents so late in the disease may not be as fruitful as in earlier disease when viral load and cytokine elaboration are less. Nevertheless, fish oils are readily available and relatively innocuous. They reduce the production of cytokines that are requisite for viral replication in lymphoid and macrophage reservoirs[28] and triglyceride levels which may be a marker of cytokine-mediated metabolic effects. Furthermore, some of the cytokine-mediated abnormalities of glucose metabolism (insulin resistance, increased gluconeogenesis) in HIV disease[29-31] are inhibited by fish oils in animal models.[32] For these reasons, it is logical and prudent to use them very early in HIV infection in the doses recommended by Endres, 18 grams daily. These can be given in divided dosage, 9 grams twice daily or 6 grams 3 times daily.

Research in early HIV infection should address the effects of fish oils on body composition, resting energy expenditure, weight gain, triglyceride levels and urinary neopterin levels, or other measures of active cytokines to confirm or refute their effects. Since total viral load can now be measured in lymph tissue by the polymerase chain reaction, studies should also include these measurements to assess efficacy. Experimental nutritional interventions in early HIV disease are summarized in Table 4.

LATENT STAGE HIV DISEASE

Viral and Immune Factors

During this stage of the illness, the HIV virus is deceptively "dormant" in lymphoid and macrophage reservoirs.[1] However, evidence of ongoing viral replication and immune system dysfunction is abundant as follicular dendritic cells trap HIV particles, and follicular lymph hyperplasia proceeds with subsequent activation of germinal center B cells and the elaboration of interferon alpha.[1]

An increase in NK cell (cytotoxic T cell) activity as an HIV-specific immune response occurs. This produces further cytokine elaboration (TNF alpha, TNF beta, INF gamma, IL-1, IL-3, IL-6). Productive infection of recirculating CD4 and F_C-receptor bearing cells continues and viral replication slowly increases, not just in the lymphoid reservoirs[1] but in epithelia[33] and reticuloendothelia[34] as well. CD4 levels fall also by apoptosis.[1] More rapid progression may occur by syncytia formation.[1]

Metabolic and Nutritional Changes

The subtle metabolic changes seen in the early stage of infection become more evident. Weight loss and regain cycles are more frequent. Futile lipid cycling results from an increased metabolic expenditure required to actively synthesize VLDL triglyceride in the liver, store it in adipose tissue, and not oxidize the fatty acids in the liver produced by cytokine-activation of hormone-sensitive lipase in fat cells.[35] Concurrently, gluconeogenesis is increased somewhat,[31] and lean body mass is further reduced. The patient undergoes a greater increase in REE and fatigue becomes a major clinical problem, as the augmented REE is used to promote futile cycling.[35,36] These, as Hellerstein has described, are the "abnormalities of cellular metabolism that direct nutrients from lean tissue into fat stores or energy wasteful futile cycles."[31]

As epithelia become invaded[33] mucosal dysfunction occurs such that brush border enzyme elaboration becomes disrupted.[37-39] Subtle deficiencies in pancreatic exocrine function are also demonstrated.[55]

Along with fatigue, anorexia appears and, while caloric intake could be adequate to maintain weight, it is more frequently inadequate to replete lean body mass.[29,40,41]

TABLE 4
Experimental Nutritional
Intervention in Early
HIV Disease

Cytokine Modulation
Fish oil supplementation
Pentoxifylline therapy
Thalidomide therapy

Nutritional Interventions

At this stage the Nutritional Task Force on AIDS[22] and the Physicians Association for AIDS Care (PAAC)[21] have established that enteral feeding supplementation be initiated with an intact formula (Table 5). If the patient can achieve an extra 360 to 550 kcal intake with an extra 15 to 20 gram protein intake daily, along with supplemental vitamins (especially vitamin A), then weight can be repleted. Again, a recent study by Hellerstein's group has shown that lean body mass was not changed in a group of HIV patients with this approach.[24]

It has been established that nutritional counseling alone has achieved weight gain, while Hellerstein has demonstrated that an appropriate refeeding schedule can achieve intakes as high as 3,000 kcal daily per patient.[24] Continued close work with the nutritional support dietician is mandatory at this stage. Again, if body composition analyses and calorimetry are available, then one could determine the responsiveness to the intervention and become more aggressive with therapy.

If augmented intake cannot be achieved, or if any weight loss continues despite enteral supplementation, then assessment of mucosal absorptive function must be undertaken. A search for the presence of either primary HIV enteropathy[39] or secondary infection of the gut[42] should be started, as therapy for secondary or opportunistic infections will resort in lean body mass repletion (Figure 1).[43]

Endocrine abnormalities may be present[44] and tests of adrenal cortical and thyroid function must be performed. Corticosteroid replacement therapy may be necessary. However, the thyroid abnormalities are cytokine driven and differ from those of the euthyroid sick state seen in sepsis, other infections or cancer,[45] and it is unclear at this time about how to intervene.

Assessment of mucosal and pancreatic function can be easily achieved by utilizing the d-xylose absorption and fecal fat excretion tests. Quantitation of such dysfunction will allow for tailored therapy. Re-testing after a period of intervention will aid in determining future nutritional support (Figure 2), such that a return to simpler enteral methods or the application of TPN may be substantiated.

TABLE 5
Enteral Supplementation Therapy in Latent Stage
HIV Disease

Intact formulas (Ensure, Nutren, Replete, Ensure with fiber,
Meritene, Citrisource, Resource, Fibersource)

Special Formulas
  Fat malabsorption: Lipisorb, Isosource
  Immune modulating: Immun-Aid, Alitraq, Impact
  Elemental: Opti, Vital HN, Peptamen, Reabilan

Pancreatic enzymes for Steatorrhea >10–20 grams daily
(Figure 2)

If mucosal absorption is normal or only minimally impaired, and anorexia persists, then gastrostomy or jejunostomy tube placement with formula infusion are prudent.[22] Patients accept these tubes more readily than nasoenteric tubes as they can be hidden from view. Kotler has demonstrated positive energy balance and an increase in lean body mass in a subset of these patients.[77] If weight gain is not achieved with an intact formula, then a peptide-based formula, perhaps with supplemental arginine, medium chain triglycerides and glutamine can be tried (Table 5) (Figure 2). The addition of pancreatic enzymes is prudent if significant steatorrhea (>10–20 g/d) is present.

It must be reemphasized that nutritional repletion alone, even at this stage of disease, does not replete lean body mass in most patients, although there may be a subset of patients, yet to be defined, that do have an increase.[40,77]

This has led to other strategies to improve LBM. The use of recombinant human growth hormone has been explored in a preliminary fashion by Hellerstein et al.[46] and Krentz[47] and show some promise in improving lean body mass (LBM), performance characteristics and lowering urea nitrogen excretion.[47] However, in Krentz's limited study, continued treatment beyond six weeks showed a loss of these favorable changes and a return to increased total body water and total body fat mass.[47]

Efforts to stimulate the appetite with mesgesterol acetate[48] have led to an increase mainly of fat weight. Studies with delta 9 THC (Marinol) have shown weight gain, but body composition analyses were not studied.[49]

Again, a multiple strategy may be needed at this stage (Table 6) that provides adequate nutrients along with pharmacologic interventions aimed at modifying the metabolic abnormalities driven by cytokines. Research in this area is needed to support such strategies.

Promise lies in newer anti-retroviral convergent therapies that exhaust the HIV virus by overwhelming viral reverse transcriptase[50] and lower viral load by vaccine and other methods. When viral load is reduced, cytokine production is reduced (decreased interferon after AZT therapy)[51] and many of the metabolic

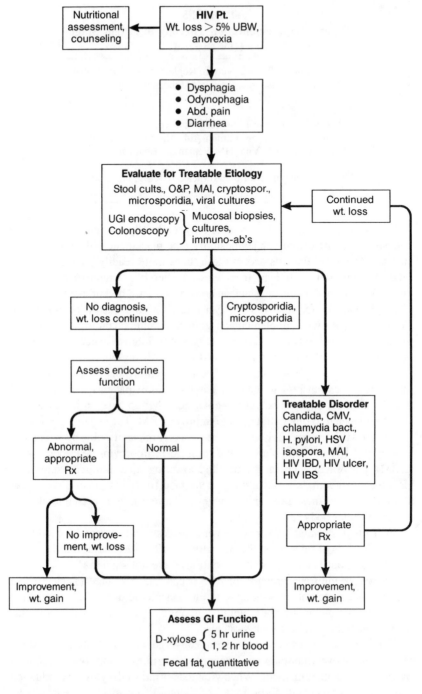

Figure 1.   Evaluation of wasting in HIV disease.

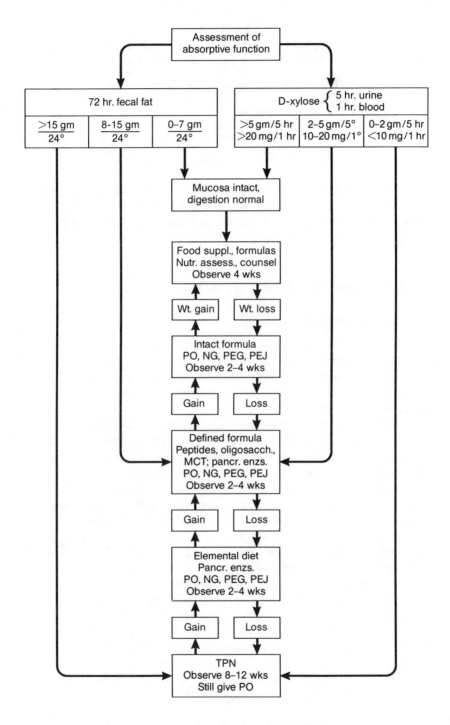

Figure 2. Nutritional therapy of wasting in HIV disease.

TABLE 6
Multiple Strategies For Optimal Nutritional
Intervention in Latent HIV Disease

- Tailor dietary supplementation to functional state of GI tract (Figure 2, Table 5)
- ? Add growth hormone
- Add Megesterol,? Dronabinol
- Continue or initiate anti-cytokine therapy (Table 4)
- Exercise

abnormalities (triglycerides lowered to normal) improve. Certainly, the advent of reverse transcriptase inhibitors has extended this period of latency before clinical AIDS develops, but has not reduced overall mortality.[52] This provides still greater opportunity for nutritional therapies and research.

LATE STAGE HIV DISEASE: AIDS

Viral and Immune Factors

As viral replication continues in lymphoid and macrophage reservoirs, destruction of the follicular dendritic cells in the germinal center occurs, leading to lymphocyte depletion and massive viremia, from "spill over" as the antigen presentation FDC cells degenerate in an "accelerated mode" related to a lack of cytotoxic activity.[1]

Cytokine activity continues but is now fueled by the appearance of opportunistic pathogens.[56] Opportunistic or secondary infections abound and account for most of the digestive and nutritional malfunction. Cytokine production is now mainly fueled by the reactions to these organisms[56] and continue to amplify the metabolic disturbances. Infections and tumors produce malabsorption, diarrhea, intestinal obstruction, hemorrhage, pancreatitis and abdominal pain. Liver dysfunction from invasion by *Mycobacterium avium intracellulare,* cytomegalovirus, *cryptosporidia* and *cryptococcus* occurs.[57] Biliary tract disease produces severe pain, pruritus, fever, nausea and vomiting.[57] Anti-retroviral, antibiotic, anti-fungal, antiviral and chemotherapy lead to oral lesions, dysgeusia, nausea, vomiting, diarrhea, pancreatitis and abdominal pain.

Futile cycling described but more subtle in earlier disease now becomes a major factor. Numerous metabolic disturbances occur (Table 7). The patient becomes further exhausted and depressed. As the patient starves from poor intake, abdominal pain, nausea, vomiting, diarrhea and malabsorption, the metabolic compensation usually present — a decrease in energy expenditure — is replaced instead by a dramatic increase in energy expenditure fueled by the cytokine-driven futile metabolic cycling.[40]

TABLE 7
Metabolic Disturbances in
Late HIV: AIDS

Increased *de novo* lipogenesis
Increased hepatic gluconeogenesis
Increased free fatty acid appearance
Increased whole body fat oxidation
Increased endogenous cholesterogenesis
Increased REE
Decreased nitrogen balance

After Hellerstein, M.K., *Nutrition and HIV/ AIDS*, Vol. 1, PAAC Publishing, Chicago, 1992, 17.

Nutritional Intervention in Late HIV Disease: AIDS

It is not until this very advanced stage that too many clinicians finally appreciate that there are nutritional problems with the patient. Body weight falls below its usual pre-illness weight by 20%, and lean body mass simultaneously loses proportionately more by another 10–15%.[3] This has been established ever since the early stage of the infection when weight was regained but was mostly water and fat rather than lean body mass. Kotler has shown that there are about 100 days of survival left for these patients regardless of the AIDS-associated process. Death occurs by malnutrition.[3] Despite the clinically obvious malnutrition, many clinicians persist in trying enteral and vitamin supplementation. When this strategy, coupled with symptomatic therapy for nausea, appetite stimulation, pain and diarrhea fails, the patient becomes moribund.

Intravenous nutritional support (TPN) is most commonly given at this late stage of the illness. Studies have shown that weight, serum albumin and sense of well-being improve with this expensive technique.[58] By and large, it is generally well-tolerated, producing no greater complication rates than are seen in cancer patients similarly treated.[58]

However, most of the weight is regained as fat,[59] although some studies suggest a subset of patients that have some repletion of lean body mass, but at a much lower rate than fat repletion.[59] If treatable opportunistic pathogens coexist, then TPN or any supply of nutrients can aid in the repletion of body weight,[43] but not commonly at this late stage.[59] Controlled trials, currently in progress, will measure the effects of TPN on nutritional parameters, clinical course and survival, incidence of complications and total cost of care. Entry criteria for these trial specify clearly that patients need only be HIV positive and to have either failed aggressive enteral nutritional support and dietary intervention or to have documented malabsorption. The trials thus propose an earlier intervention for TPN than is now practiced. These should do much to resolve controversy over the use of TPN.[60]

Energy expenditure in stressed, ICU patients not dissimilar to AIDS patients with secondary or opportunistic infection is often overestimated by the Harris-Benedict equation, and overfeeding of these patients occurs.[61] Excessive body fat is produced and a diminished ability to prevent lean body mass wasting has been demonstrated.[62] Optimally, indirect calorimetry should be used to estimate energy requirements with energy supply given with an activity factor of 15% added for full ambulation.

In this setting, the major energy source used is dextrose. One should not exceed 5 mg of dextrose/kg body weight/minute as this is the maximum oxidation rate of glucose in humans.[63] One would not want to force the patient to store energy as fat by giving more than this. Many of the studies that showed fat storage from TPN did not specify the rate of glucose administration, leading one to speculate that these patients may have been overfed with dextrose.[59]

Lipids should be provided daily. From the earliest days of TPN, when 25% dextrose and 5% amino acids were given daily, and lipids were provided twice to three times a week to all patients, reimbursement policy by Medicare has established that lipids be given three times a week. This absurdity has led to TPN prescribing practices by physicians causing days when all of the non-protein energy is carbohydrates to excess 4 days a week, and excessive non-protein energy is given the other 3 days a week. If we would all eat this way, obesity would be even more common than is currently present. Studies show that a 70% dextrose to 30% lipid mixed energy supply to surgical patients leads to an improved repletion of lean body mass, with less fat and water accumulation than in patients given dextrose as energy and lipid to supply adequate essential fatty acids.[64,65] Again, careful review of the previously published studies[59] shows no specification for the lipid percentage provided to these patients. Furthermore, in Kotler's study, patients with anorexia or malabsorption received 43.4 ± 6.2 kcal/kg body weight when the goal was to give a maximum of 35 kcal/kg body weight. Yet many of these patients gained an increase of lean body mass as measured by total body potassium.[59] The group with systemic, opportunistic infections received 50.5 ± 4.2 kcal/kg body weight and mostly regained fat as one would predict, but 2 of 7 in that group still had a gain in lean body mass. This suggests that all of these patients may have been overfed leading to fat storage. As Baker[61] and Streat[62] have shown, REE in ICU patients with similar infections is only modestly increased.

To optimize nutritional support, calories should best be adjusted by indirect calorimetry. Lacking the availability of this technique, total caloric load in these patients should not exceed 35 kcal/kg body weight per 24 hours. When 25 AIDS patients were given home TPN under the guidance of indirect calorimetry, more than 60% regained significant lean body mass versus body fat by anthropometry.[67] In the ongoing studies of TPN in HIV disease, both bioelectric impedance and indirect calorimetry are being utilized to optimize support. This may answer many of the questions about the nature of weight gain from TPN in AIDS patients.

Concern exists with the use of lipids in immunosuppressed patients. Adverse effects on platelet, lymphocyte, fixed macrophage and PMN leukocyte function have all been demonstrated.[68] However, if the rate of lipid infusion is kept to 0.03 to 0.05 g/kg.hour then none of these effects are seen. Indeed, one can give up to 0.11 gm/kg.hour safely to most patients. Lipids should be infused over at least 12 to 14 hours and should compromise about 30% of nonprotein calories.[68]

Singer and Askanazi have shown excellent tolerance of IV lipid when given to AIDS patients at 50% of nonprotein calories.[58] The phospholipid portion of intravenous lipids is similar to compound AL-721[76] found to have anti-HIV properties. The excellent results in their experience with TPN in AIDS may be partially attributable to this property of phospholipid. Further studies using just this phospholipid as the lipid component of TPN are currently in progress.[67]

Protein requirements are similar to those of HIV negative patients given TPN.[59] If the therapeutic goal is to maintain current weight or lean body mass, then 1.0 to 1.2 g protein as amino acid per kg daily are required. If repletion is desired, then 1.5 to 1.7 g/kg/d are given (Table 8).

Experimental Approaches

The use of nutrient components for their pharmacologic properties rather than for their nutrient effects warrants investigation. Cytokine modification by feeding 18 g/d of fish oil for 10 weeks was attempted by Hellerstein in 20 AIDS patients with substantial weight loss. These reduced triglycerides, but had no effect on cytokine production *in vitro* from endotoxin-simulated peripheral blood mononuclear cells.[29] It is possible, in addition to the already weak cytokine production by these peripheral blood mononuclear cells from malnourished patients, and fish oils being weak anti-cytokine agents, that adequate levels of fish oil were not achieved in these patients who likely had malabsorption. Intravenous fish oil preparations are available and have been safely used in septic patients and patients with cystic fibrosis.[68] Using fish oil as a lipid caloric source is an attractive research study.

Medium chain triglycerides given intravenously have reduced fatty liver and have served as an energy source to fixed macrophages.[66,69] These should be investigated. The aforementioned phospholipid with anti-HIV virucidal effects is being studied and possibly requires reformulation. It would be attractive to study a "superlipid" molecule consisting of fish oil, medium chain triglyceride and phospholipid in AIDS patients requiring TPN.

Free glutamine has been unstable in TPN amino acid solutions, and has hence been withheld from them. Glutamine is a major energy source for lymphocytes[71] and macrophages,[72] as well as for the small intestinal and, to a lesser extent, colonic mucosae.[73] It is the major free amino acid in muscle and is rapidly depleted during surgical and infectious stress. Glutamine given to

TABLE 8
Energy and Protein Requirements for TPN in HIV Disease

Energy by calorimetry with addition of 15% for activity, or, 35 kcal/kg daily
  maximum
Do not exceed 5 mg/kg.min of dextrose daily (70% of non- protein kcal)
Provide daily lipids and do not exceed 0.11 g/kg·h (30% of non-protein kcal)
For maintenance give 1.0–1.2 g amino acids/kg·24 h
For repletion give 1.5–1.7 g amino acids/kg·24 h

animals and humans as a stabilized amino acid or as glutamine-alanyl dipeptide
has restored intestinal function in ICU patients[74] and replaced muscle glutamine
with improved nitrogen balance in surgical patients.[74] Glutamine-supplemented
TPN reduced opportunistic infections, improved nutritional parameters and
shortened hospital stay in bone marrow transplant patients.[78] Studies with IV
glutamine are being conducted in AIDS patients to see if gastrointestinal
function, immune function and reduced numbers of opportunistic infection and
improved survival occur.[60]

Concurrent intervention with more potent anti-cytokine medications such
as pentoxifylline can also be studied. Once much of the ongoing research is
completed, we may have more powerful and optimal nutritional intervention
for patients with HIV disease. The future for nutritional intervention indeed
looks bright and could be a major contributor in reducing co-morbidity and
extending life in this dreadful epidemic.

REFERENCES

1. Pantaleo G, Graziosi C, Fauci AS: The immunopathogenesis of human immunodeficiency virus infection. *New Engl J Med* 1993;328:327-335.
2. Chlebowski RT, Grosvenor MD, Bernhard NH, et al: Nutritional status, gastrointestinal dysfunction, and survival in patients with AIDS. *Am J Gastroenterol*
   1989;84:1288-93.
3. Kotler DP, Tierney AR, Wang J, and Pierson RN, Jr.: The magnitude of body cell
   mass depletion determines the timing of death from wasting in AIDS. *Am J Clin
   Nutr* 1989;50:444-447.
4. Levy JA: Human immunodeficiency viruses and the pathogenesis of AIDS.
   JAMA 1989;261:2297-3006.
5. Tersmatte M, Gruters RA, DeWolf F, et al: Evidence for a role of virulent human
   immunodeficiency virus (HIV) variants in the pathogenesis of acquired immunodeficiency syndrome. Studies on sequential HIV isolates. *J Virol* 1989;63:2118-
   2225.
6. Schnittman SM: Viral burden in human immunodeficiency virus type 1 (HIV-1)
   infection. In Immunopathogenic mechanisms in human immunodeficiency virus
   (HIV) infection. *Ann Intern Med* 1991;114:678-693.

7. Lifson AR, Buchbinder SP, Sheppard HW, et al: Long term human immunodeficiency virus infection in asymptomatic homosexual and bisexual men with normal CD4+ lymphocyte counts: Immunologic and virologic characteristics. *J Infect Dis* 1991;163:959-65.

8. Poli G and Fauci AS: The effect of cytokines in pharmacologic agents on chronic HIV infection. *AIDS Res Hum Retroviruses* 1992;8:191-7.

9. Molina JM, Scadden DR, Byrn R, et al: Production of tumor necrosis factor and interleukin-1 beta by monocytic cells infected with human immunodeficiency virus. *J Clin Invest* 1989;84:733-7.

10. Rosenberg ZF and Fauci AS: Immunopathogenetic mechanisms of HIV infection: Cytokine induction of HIV expression. *Immunol Today* 1990;11:176-80.

11. Biswas P, Poli G, Kinter AL, et al.: Interferon-gamma induces the expression of human immunodeficiency virus in persistently infected promonocytic cells (U1) and redirects the production of virions to intracytoplasmic vacuoles in phorbol myristate acetate-differentiated U1 cells. *J Exp Med* 1992;176:739-50.

12. Fuchs D, Hausen A, Reibenneger G, et al: Interferon-gamma concentrations are increased in serra from individuals with human immunodeficiency virus type 1. *J Acquir Immune Defic Syndr* 1989;2:158-162.

13. Lazdins JK, Klimkait T, Woods-Cook K, et al: In vitro effect of transforming growth factor beta on progression of HIV-1 infection in primary mononuclear phagocytes. *J Immunol* 1991;147:1201-7.

14. Poli G, Kinter AL, Justement JS, et al: Transforming growth factor beta suppresses human immunodeficiency virus expression and replication in infected cells of the monocyte/macrophage lineage. *J Exp Med* 1991;173:589-97.

15. Graziosi Z, Pantaleo G, Kotler DP, and Fauci AS: Dissociation between HIV expression in the peripheral blood versus lymphoid organs of the same patients. *Clin Res* 1992;40:333A (Abstract).

16. Hommes MJT, Romjin JA, Edert E, et al: Resting energy expenditure and substrate oxidation in human immunodeficiency virus (HIV)-infected asymptomatic men. HIV affects host metabolism in the early asymptomatic stage. *Am J Clin Nutr* 1991;54:311-15.

17. Ott M, Lembcke B, Fischer H, et al: Early changes of body composition in human immunodeficiency virus-infected patients. Tetra polar body impedance analysis indicates significant malnutrition. *Am J Clin Nutr* 1993;57:15-19.

18. Hommes MJT, Romijn JA, Endert E, et al: Basal fuel homeostasis in symptomatic human immunodeficiency virus infection. *Clin Sci* 1991;80:359-65.

19. Zangerle R, Reibnegger G, Wachter H, and Fuchs D: Weight loss in HIV-1 infection is associated with immune activation. *AIDS* 1993;7:175-181.

20. Grunfeld C, Kotler DP, Hamadeh R, et al: Hypertriglyceridemia in the acquired immunodeficiency syndrome. *Am J Med* 1989;86:27-31.

21. Physicians Association for AIDS Care. *Nutrition and HIV/AIDS*. Proceedings of the 1992 International Symposium on Nutrition and HIV/AIDS Including the Nutrition Algorithm and Nutrition Initiative of the Physicians Association for AIDS Care. Volume 1, 1992. PAAC Publishing Inc., Chicago.

22. Task force on nutritional support in AIDS (1989). Guidelines for nutritional support in AIDS. *Nutrition* 1989;5:39-45.

23. Rigsby LW, Dishman RK, Jackson AW, et al: Exercise for seropositives for the human immunodeficiency virus-1. *Med Sci Sports Exerc* 1992;24:6-12.

24. Hellerstein MK, Hoh R, Neese R, et al: Effects of nutritional supplements of different composition on nutritional status and gut histology in HIV wasting. Metabolic abnormalities for prediction of nutrient unresponsivity. Presentation. AIDS VIII. International Conference on AIDS. Amsterdam, 1992.

25. Mathys P, Dukmans R, Proost P, et al: Severe cachexia in mice inoculated with interferon-gamma producing tumor cells. *Int J Cancer* 1991;49:77-82.

26. Strieter RM, Remic DG, Ward PA, et al: Cellular and molecular regulation of tumor necrosis factor alpha production by pentoxifylline. *Biochem Biophys Res Commun* 1988;155:1230-1238.

27. Sampiano EP, Samo EN, Galilly R, et al: Thalidomide selectively inhibits tumor necrosis factor alpha production by stimulating human monocytes. *J Exp Med* 1991;173-699-703.

28. Endres S, Ghorbani R, Kelley VE, et al: The effect of dietary supplementation with n-3 polyunsaturated fatty acids on the synthesis of interleukin-1 and tumor necrosis factor by mononuclear cells. *N Engl J Med* 1989;320:265-71.

29. Hellerstein MK, Woo K, Kempfer S, et al: Nutritional and metabolic effects of dietary n-3 fatty acid supplementation in men with weight loss associated with the acquired immunodeficiency syndrome. Submitted, 1992.

30. Kinsella JE, Lokesh B, Broughton S, et al: Dietary polyunsaturated fatty acids and eicosanoids. Potential effects on the modulation of inflammatory and immune cells. An overview. *Nutrition* 1990;6:24-44.

31. Hellerstein MK: Pathophysiology of lean body wasting and nutrient unresponsiveness in HIV/AIDS: Therapeutic implications. *Nutrition and HIV/AIDS,* Vol. 1, 1992, pg 17-25, PAAC Publishing, Inc., Chicago.

32. Ling PR, Istfan N, Colln E, and Bistrin BR: Effects of fish oil on glucose metabolism in the interleukin-1 alpha-treated rat. *Metabolism* 1993;42:81-85.

33. Heise C, Dandekar S, Kumar P, et al: Human immunodeficiency virus infection of enterocytes and mononuclear cells in human jejunal mucosa. *Gastroenterology* 1991;100:1521-7.

34. Buhl R, Jaffe HA, Holroyd KJ, et al: Activation of alveolar macrophages in asymptomatic HIV-infected individuals. *J Immun* 1993;150:1019-28.

35. Grunfeld C and Feingold KR: Metabolic disturbances and wasting in the acquired immunodeficiency syndrome. *N Engl J Med* 1992;327:329-337.

36. Grunfeld C, Pang M, Doerrler W, et al: Lipids, lipoproteins, triglyceride clearance and cytokines in human immunodeficiency virus infection and the acquired immunodeficiency syndrome. *J Clin Endocrinol Metab* 1992;74:1045-1052.

37. Yolken RH, Hart W, Oung I, et al: Gastrointestinal dysfunction and disaccharide intolerance in children infected with human immunodeficiency virus. *J Pediatr* 1991;118:359-63.

38. DaCunha-Ferreira R, Forsythe, Richmond PI, et al: Changes in the rate of crytoepithelial cell proliferation and mucosal morphology induced by a T-cell response in human small intestine. *Gastroenterology* 1990;98:1255-63.

39. Ullrich R, Zeitz M, Heise W, et al: Small intestinal structure and function in patients infected with human immunodeficiency virus (HIV): Evidence for HIV enteropathy. *Ann Intern Med* 1989;111:15-21.

40. Grunfeld C, Pang M, Shimizu L: Resting energy expenditure, caloric intake and short-term weight change in human immunodeficiency virus infection and the acquired immunodeficiency syndrome. *Am J Clin Nutr* 1992;55:455-60.

41. Melchior JC, Salmon D, Rigaud D, et al: Resting energy expenditure is increased in stable, malnourished HIV-infected patients. *Am J Clin Nutr* 1991;53:437-41.

42. Kotler DP, Francisco A, Clayton F, et al: Small intestinal injury and parasitic disease in the acquired immunodeficiency syndrome (AIDS). *Ann Intern Med* 1990;113:444-449

43. Kotler DP, Tierney AR, Attilio P, et al: Body mass repletion during gangciclovir therapy for cytomegalovirus colitis. *Am J Clin Nutr* 1989;49:237-41.

44. Lopresti JS, Fried JC, Spencer ZA, and Nicoloff JT: Unique alteration of thyroid hormone indices in the acquired immune deficiency syndrome. *Ann Intern Med* 1989;110:970-75.

45. Hommes MJT, Romjin JA, Endert E, et al: Hypothyroid-like regulation of the pituitary-thyroid axis in stable HIV-infection. *Metabolism* (in press) 1993.

46. Mulligan K, Grunfeld C, Hellerstein M, and Schambelan M: Growth hormone treatment of HIV-associated catabolism. *FASEB J* 1992;6:A1942 (abstr.).

47. Krentz AJ, Koster FT, Crist DM, et al: Anthropometric, metabolic and immunological effects of recombinant growth hormone in AIDS and AIDS-related complex. *J Acq Immun Def Syndr* 1993;6:245-51.

48. Vonroenn JH, Murphy RL, Weber KM, et al: Magesterol acetate for treatment of cachexia associated with human immunodeficiency virus infection. *Ann Intern Med* 1988;109:840-841.

49. Gorter R and Wolberding P: Effect of delta-4-CHC (dronabinol) on weight gain in AIDS. Personal communication.

50. Chow YK, Hirsch MS, Merrill DP, et al: Use of evolutional limitations of HIV-1 multidrug resistance to optimize therapy. *Nature* 1993;361:650-53.

51. Mildvan D, Machado SG, Wilets I, and Grossberg SE: Endogenous interferon on triglyceride concentrations to assess response to zidovudine in AIDS and advanced AIDS-related complex. *Lancet* 1992;1:453-456.

52. Kahn JO, et al: Controlled trial comparing continued zidovudine and didanosine in human immunodeficiency virus infection. *New Engl J Med* 1992;327;581-8.

53. Kotler D, Refka S, Borich A, and Kronin W: Detection, localization and quantitation of HIV-associated antigens in intestinal biopsies for patients with HIV. *Am J Pathol* 1991;139:823-830.

54. Kotler DP, Reka A, Hecker L, et al: Treatment of HIV-associated mucosal inflammation with oral 5-ASA. *Gastroenterology* 1991;100:A589.

55. Pezzilli R, Gullo L, Ricchi E, et al: Serum pancreatic enzymes in HIV-sero positive patients. *Dig Dis Sci* 1992;37:286-288.

56. Lahdevirta J, Maury CPJ, Teppo AM, and Repo H: Elevated levels of circulating cachectin/tumor necrosis factor in patients with acquired immunodeficiency syndrome. *Am J Med* 1988;85:289-291.

57. Margulis SJ and Jacobson IM: Hepatobiliary and pancreatic manifestations of AIDS. *Sem Gastrointest Dis* 1991;2:49-61.

58. Singer P, Rothkopf MM, Kvetan V, et al: Nutrition, the gastrointestinal tract and the acquired immune deficiency syndrome: Facts and perspectives. *Clin Nutr* 1989;8:281-287.

59. Kotler DP, Tierney AR, Culpepper-Morgan JA, et al: Effect of home parenteral nutrition on body composition in patients with acquired immunodeficiency syndrome. *J Parent Ent Nutr* 1990;14:454-458.

60. Aron JM, Askanazi J, and Bellos N: The effects of standard and glutamine supplemented total parenteral nutrition in HIV disease. A prospective, randomized, double blind controlled trial. In progress, 1993.

61. Baker JP, Detsky AS, Stewart S, et al: Randomized trial of total parenteral nutrition in critically ill patients: Metabolic effects of varying glucose-lipid ratios as the energy source. *Gastroenterology* 1984;87:53-9.

62. Streat SJ, Bettoe AH, and Hill GL: Aggressive nutritional support does not prevent protein loss despite fat gain in septic intensive care patients. *J Trauma* 1987;27:262-266.

63. Wolfe RR, O'Donnell DF, Stone MD, et al: Investigation of factors determining the optimal glucose infusion rate in total parenteral nutrition. *Metabolism* 1980;29:892-900.

64. Macfie J, Smith RC, and Hill GL: Glucose or fat as a nonprotein energy source: A controlled clinical trial in gastroenterological patients requiring intravenous nutrition. *Gastroenterology* 1981;80:103-7.

65. Bresson JL, Bader B, Rocchiccioli F, et al: Protein-metabolism kinetics and energy substrate utilization in infants fed parenteral solutions with different glucose-fat ratios. *Am J Clin Nutr* 1991;54:370-6.

66. Quigley EMM, Marsh MN, Shaffer JL, and Markin RS: Hepatobiliary complications of total parenteral nutrition. *Gastroenterology* 1993;104:286-301.

67. Aron JM and Askanazi J: Open label trial of intravenous AL-721 as a component of total parenteral nutrition in AIDS patients. In progress, 1993.

68. Miles JM: Intravenous fat emulsions in nutritional support. *Curr Opin Gastroenterol* 1991;7:306-311.

69. Manner T, Katz DP, Askanazi J, et al: Parenteral fish oil administration in patients with cystic fibrosis. *J Parent Ent Nutr* 1993;17:A8.

70. Kotter R, Johnson RC, Young SK, et al: Competitive effects of long chain triglyceride emulsion on the metabolism of medium chain triglyceride emulsions. *Am J Clin Nutr* 1989;50:794-800.

71. Ardawi MSM and Newsholme EA: Intracellular localization and properties of phosphate-dependent glutaminase of rat mesenteric lymph nodes. *Biochem J* 1984;217:289.

72. Ardawi MSM and Newsholme EA: Glutamine metabolism in lymphocytes of the rat. *Biochem J* 1983;212:835.

73. Newsholme P, Gordon S, and Newsholme EA: Rates of utilization in fates of glucose, glutamine, pyruvate, fatty acids and ketone bodies of muscle macrophages. *Biochem J* 1987;242:631.

74. Windmueller HG and Spaeth AE: Uptake and metabolism of plasma glutamine by the small intestine. *J Biol Chem* 1979;249:5070.

75. Stehle P, Mertes N, Puchstein N, et al: Effect of parenteral glutamine peptide supplements on muscle glutamine loss and nitrogen balance after major surgery. *Lancet* 1989;2:31.

76. Fields-Newman C: The role of nutritional assessments and nutritional plans in the management of HIV/AIDS. An overview of the PAAC Nutrition Initiative. *Nutrition and HIV/AIDS*. 1992;1:57-106. PAAC Publishing, Inc., Chicago.

77. Sarin PS, Gallo RC, Scheer DI, et al: Effects of a novel compound (AL-721) on HTLV III infectivity *in vitro*. *New Engl J Med* 1985;313:1289-90.
78. Ferraro R, Kotler DP, Cuff P, et al: Effect of enteral nutritional therapy on body cell mass in AIDS. *Clin Res* 1989;37:330A (abstr.).
79. Ziegler TR, Young LS, Benfell K, et al: Clinical and metabolic efficacy of glutamine-supplemented parenteral nutrition after bone marrow transplantation. A randomized, double-blind, controlled study. *Ann Intern Med* 1992;116:821-6.

Chapter

15

AIDS and Traditional Food Therapies

Jacquelyn H. Flaskerud

BACKGROUND

Foods and herbs have traditionally been used as home remedies or therapies to prevent, treat and cure illness. This tradition continues currently with an emphasis on health and a healthy lifestyle, including what a person eats. Although many of the current uses of nutrition to promote health are considered modern in thought, many also are grounded in traditional beliefs and theories about the uses of food to promote health and prevent illness. This chapter will focus on the traditional beliefs of Americans about nutrition and specifically on beliefs about the use of foods and herbs to prevent and treat AIDS.

EXPERIMENTAL FINDINGS

In three separate studies of the health beliefs of low income white, black, and Latina women about AIDS, Flaskerud and colleagues[7-9] found that each of the three groups had integrated its explanation of AIDS into a preexisting conceptualization of illness: its cause, prevention and treatment. Although each group was aware of the current biomedical and public health explanations of AIDS, each also had a well developed lay conceptualization of health and illness. AIDS had been incorporated within this conceptualization across the range of prevention, cause and treatment.

A second finding by Flaskerud and associates was that the use of foods and herbs as therapy was prominent in all conceptualizations of treatment. Treatment for AIDS with the use of foods and herbs was recommended by each of

0-8493-7842-7/94/$0.00+$.50
© 1994 by CRC Press, Inc.

the three ethnic groups. Specific foods and herbal remedies were prescribed to treat a variety of symptoms and diseases associated with AIDS. These remedies were congruent with the traditional beliefs of each group about specific foods and fit with their traditional theories of illness.

Other investigators also have documented the integration of new treatments into existing conceptualizations of treatment. Logan[17] notes that the cognitive system underlying various classifications remains constant. Consequently, medicines not yet known to a patient will, upon introduction, be classified into an existing system. In the particular case of which Logan was writing (hot/cold theory), medicines will be classified as having hot or cold properties. Similarly, new illnesses will also be classified as hot or cold.[17] Harwood[14] takes a similar view noting that hot-cold theory is very functional and orders a great deal of health behavior, thus reinforcing and validating its use for groups who practice it. The assumption can be made that other theories and classifications of illness, treatment and foods serve similar functions for persons of other cultures. A review of the major traditional theories of illness provides a background for the longstanding use of food as therapy and for understanding nutrition beliefs about AIDS and its treatment.

TRADITIONAL THEORIES AND NUTRITION

Americans are pluralistic in ethnicity and culture, yet many beliefs about the uses of food as therapy have common explanations for the relationship between nutrition and health, how specific foods work and when to use them. Common themes are that foods build up the body's strength, ward off illness, fight illness and cleanse the body of impurities. The notion of balance in the diet is prevalent in most traditional beliefs.

Hot/Cold Theory and Nutrition

The Hippocratic theory of humoral medicine guides the traditional use of food as therapy for a wide range of Americans. Humoral science originated in the Mediterranean area, spread to the Arabic world and was brought to the Americas and parts of Asia during Spanish colonialism.[6,13,17] The influences of Hippocratic medicine are present in the self-care beliefs and practices of Latin Americans and of Southeast Asian and Filipino immigrants to the United States.[1,12,20] Humoral science also underlies the beliefs of Indian traditional medicine which is based on Ayurvedic principles.[22] Americans who had their origins in any of the Mediterranean countries, Middle East and Arab countries, some Asian countries, India and Latin America share similar beliefs about the properties and use of food as medicine. Properties of hot and cold are assigned to foods and to illnesses. The concept of balance is used to prescribe particular foods for specific illnesses depending on the property of the illness.

Humoral nutritional beliefs encompass prevention, cause of illness, cure and nourishment. Cultural groups which incorporate traditional hot-cold theory into their medical beliefs classify food stuffs as hot or cold, with a number of foods classified as neutral, having neither hot nor cold qualities.[6,14,21] These properties of food have functional significance and order health behaviors. Health is maintained by diet as long as a person's equilibrium of hot and cold is not upset. Balance in hot and cold is important to prevent disease and to prolong life, as is moderation in ingestion of food and drink. Eating too many hot foods (or cold) can create a reaction that may be alleviated by eating foods of the opposite property.

Warm foods are believed to be more easily digested than cold foods.[6] A diet consisting of entirely warm foods will not necessarily make a healthy person sick; however, a diet consisting of cold foods alone will cause illness in even the healthiest person. Consuming too many cold foods without an equivalent amount of hot food can cause sickness as can the opposite. Certain physical problems resulting from the overconsumption of, for instance, hot foods can be ameliorated by specific cold foods or drink.[21]

Sickness is not caused solely by the overconsumption of one category of food. There are other causal agents of illness as well, and it is the problem of disease that gives the classification of foodstuffs their ultimate importance in cure. In general, diseases or symptoms classified as hot are treated with foods, medicines and herbs that are cold and vice versa. For purposes of the discussion of AIDS and traditional food therapies later in this chapter, certain specific aspects of the use of food to treat illness are emphasized here. Blood is the primary source of life in the body, of bodily strength and of warmth. Any condition that results in loss of blood also results in weakness and an excess of cold. Such a condition is treated by strengthening foods with hot properties, by ingesting foods containing blood or blood itself, and often by ingesting foods, liquor, wine and spices with darker colors.[6,11,17,20,21]

Fever, stomachache, diarrhea, constipation, nausea, rashes and pustules are often attributed to impurities in the blood and an excess of heat. Cleaning the blood of impurities can be accomplished by the use of food and herbal remedies considered to have cold qualities and cleansing or purgative properties.[14,21] Many fresh fruits and vegetables and greens have cold properties. Chills, chronic cough without blood, tuberculosis and pneumonia (without blood) are all considered an excess of cold. Prescribed treatments would avoid foods, medicinal plants and modern medicines with cold properties. The practice of giving hot or cold foods in particular kinds of clinical conditions is a balancing technique for treating the condition and not necessarily for nourishment.[6,17]

Nourishment is, however, important in recovery from illness. During convalescence, the body is considered weak and proper food will strengthen the blood and the body. In addition to aiding recovery, food and drink supply nourishment to the body and strengthen resistance to disease.

Yin/Yang Theory and Nutrition

In Chinese culture the ancient philosophy of Tao proposes a balance in the universe based on elements of yin and yang. Yin is cold, darkness, earth, repose and femaleness; and yang is heat, light, heaven, movement and maleness.[19,28] Foods are also grouped as yin or yang. Food has properties such as cool, warm, cold and hot and flavors such as sweet, sour, bitter, pungent and salty.[29] A balance of yin and yang in the diet is vital to good health. Yin conditions require yang foods for treatment and vice versa in order to maintain balance within the body. Food is used as therapy in prevention of disease and cure and as well as in nourishment.[29] There is a striking resemblance between Hippocratic humoral theory and the centuries older Chinese yin-yang philosophy which might suggest a center of origin for humoral theory other than that of the Mediterranean area.[17]

Americans of Asian origin who share traditional food therapy beliefs based on yin and yang include Chinese, Filipinos and Southeast Asians.[1,4,12,20,29] Food is important for sustaining life and good health, for the basic strength of the body, and for curing and nourishment.[29] As in humoral theory, food is used as therapy to build the blood or strengthen and fortify the blood.[19,28,29] Yang foods (hot foods) are used for blood building. Yang foods include beef, eggs, garlic, liquor, peanuts, spicy foods, wine, red beans, red foods and red peppers. For blood building, pork liver is considered an important part of diet. Again, these foods are remarkably similar to those recommended for blood building in humoral therapy.

Fevers, infections, venereal disease, upset stomach and constipation are examples of yang conditions and are treated with yin foods. Yin foods include most fresh fruits and vegetables, some fish and greens.[2,19] Shivering, wasting, cancer, and diseases of the lungs are considered yin conditions and should be treated with warm foods, while salty and cold foods should be avoided.[19,29]

Natural/Supernatural Conceptualizations and Nutrition

Snow[24,25] has conceptualized the beliefs of African Americans regarding health and the cause and treatment of illness as falling into natural and unnatural categories. Natural illnesses are those that can be attributed to nature and nature's God. Unnatural illnesses are those attributed to forces of evil, witchcraft and the devil. Flaskerud and Rush[7] categorized these sources of illness as natural, i.e. those caused by their relationship with nature, and supernatural, including those caused by both good and bad supernatural forces. For purposes of this discussion, natural illnesses are reviewed, as they are the ones associated with food therapy and nutrition. Again those beliefs that will be specifically related to AIDS and nutrition later in this chapter are discussed in detail.

Natural illnesses are caused by cold, dirt or impurities, and improper diet.[25] All of these causes are related to blood and its functions, and a moderate and healthy life style that emphasizes balance. Blood is viewed as good or bad, clean or dirty, thick or thin, high or low, and sweet or bitter. These beliefs are found among African Americans, white Southerners, immigrants from Haiti, Jamaica and the Bahamas.[18,23,25,26]

Exposure to cold air, cold water and dampness causes acute illnesses associated with mucous production such as pneumonia, bronchitis, cold, flu and the development of arthritis and rheumatism. Persons whose blood is thin are more susceptible to cold. The young, old, menstruating and postpartum women have thin blood. Building up one's blood (or thickening it) through food can protect a susceptible person against cold and should be practiced by everyone in the winter.[18] Specific foods include liver, pork, beets, and wine. Also protective against cold is dressing warmly and avoiding cold air, drafts, dampness, and water.

Dirt, impurities and a general notion of germs are causes of illnesses associated with heat and exemplified by illness conditions such as fever, inflammation, skin eruptions, measles and venereal disease.[24,25] Impurities are circulated within the body through the blood and expelled from the body through menses, bowel movements, hives and rashes, and the pores. Impurities in the blood are caused by failure to bathe, dirty clothes and house, impeded menses, irregular bowel movements and sexual excess. The blood can be cleaned out by using purgatives, sulfur and molasses, greens and teas.[23,25]

Diet is believed to prevent illness, cause illness and to have curative properties.[26] Two conditions of the blood are associated with diet: (1) high blood, thick blood, sweet or too much blood; and (2) low blood, thin blood or not enough blood. High blood is caused by too many rich foods or "blood builders" in the diet. These are foods red in color: red meat (especially pork and liver), beets, grape juice, red wine, raisins and black molasses.[23,25] The blood can be thinned by white or colorless foods and bitters: lemon, vinegar, epsom salts, garlic and onions. These lower or thin out the blood through the pores, bowels and menses. Persons with low blood are treated with the blood builders referred to earlier. It is protective to build up one's blood for winter and thin it out for spring and summer.[18,23] Balance in the diet and monitoring the state of the blood is considered essential to health. Once again in this conceptualization as in hot/cold and yin/yang theories, there are notions of illness expressive of cold and heat, of the blood as signifying the body's strength and of the use of food to treat variations in the blood. The foods used to build up and thin out the blood are remarkably similar among the three conceptualizations.

Feed a Cold/Starve a Fever

Several investigators have documented the use of folk conceptualizations of illness and folk remedies among Americans of European origin.[3,5,9,15,16] These beliefs and practices have been found with both lower and middle class, and urban and rural subjects.[3,9,15] Many of the remedies endorsed may be classified as dietary substances.[5,15]

The conceptualizations of illness used by white Americans of European origin bear a striking resemblance to many of the beliefs of other American ethnic groups and a particular affinity to the beliefs of African Americans.[16,25] Helman[16] describes a folk model of illness that includes hot illnesses generally associated with fever (and with infectious diseases in biomedicine) and cold illnesses associated with chills. Cold illnesses involve the relationship between persons and nature or the natural environment, and hot illnesses involve the relationship between persons in society. Cold illnesses are caused by damp, rain, cold air, night air, drafts and climatic changes.[15,16] Such illnesses enter through the skin, ears, top of the head, back of the neck and the feet. They cause colds, flu, rheumatism, bronchitis, pneumonia, cold loose stools and cold in the kidneys.[5,15,16] Prevention of cold illnesses includes staying warm and strengthening the body with foods and tonics. Hot drinks, rest in a warm bed and ample warm foods are used to treat cold. Tonics, vitamins, cod liver oil and whiskey are also used, but the most important body strengthener is food.

Hot illnesses, fevers, hot diarrhea, boils and rashes are caused by impurities in the blood, concentrations of dirt, bugs, germs or viruses. They originate in other people (not in the natural environment), are transported by people and enter the body through orifices such as the mouth, nose, anus, urethra and vagina.[15,16] Hot illnesses are treated by restricting foods and flushing them out of the system with lots of fluids. Evidence that these hot illnesses are leaving the body are diarrhea, vomiting, sweating, expectorating, urinating, blowing nose and skin rashes. Remedies used to flush out the body or cleanse the blood are laxatives, vinegar, ginseng and garlic. Fresh fruit juices in large amounts are the most common fluids recommended for flushing out the system when fever is present.[15] Also, for a person who is feverish, reducing food intake is used concomitantly with forcing fluids.

AIDS AND TRADITIONAL FOOD THERAPIES

Flaskerud and colleagues[7-9] studied the beliefs about AIDS of white, black and Latina low income women in Los Angeles. Data were gathered through semi-structured focus group interviews. Each ethnic group was interviewed separately by data collectors who shared the ethnicity, language and gender of the subjects. Knowledge of AIDS differed among the subjects. Those who had no personal experience with the disease tended to classify it as a hot illness. AIDS was associated with venereal disease, impurities in the blood, fever, rashes and dirty practices and habits. AIDS was believed to be transmitted from one person to the next through body orifices and pores, by direct contact with an infected person or indirectly by contact with objects such as toilet seats, furniture, eating utensils, swimming pools and spas that were contaminated with body fluids (especially urine, blood, feces and salvia). The AIDS virus was believed to be excreted from the body through urine, feces, sweat, menstrual blood and sexual fluids. To cleanse the body and purify the blood of the virus, Latina women recommended eating fresh fruit, drinking wheat and barley waters, and purging the body with herbs, mineral oils, laxatives and other purgatives. Ingestion of garlic also was recommended to purify the body and blood of the AIDS virus and cleanse the stomach and bowels. Additionally, garlic was thought to protect against AIDS infection through mosquito bites. Several herbal remedies available in Mexico included teas made from roots or bark that were believed to purify the blood of both syphilis and AIDS. Finally, treatment with snake blood was recommended to drive out the poisons.

Among African American respondents, cleansing the blood and the body were considered methods of ridding the body of the AIDS virus. Food therapies recommended were epsom salts, black draught, castor oil, other laxatives, greens, astringents and acids such as vinegar and lemon, lemon tea or lemon stew. Some African American subjects had experience or had heard that persons with AIDS had fevers and large amounts of diarrhea. They believed it was a mistake to try to stop this diarrhea as it was an attempt by the body to excrete toxins. Otherwise the "toxins in your wastes would be reabsorbed by your blood stream." Keeping "your system cleaned out" was considered a method of staying healthy as well as a possible cure. Fevers with AIDS were also treated by purging the blood via laxatives, greens and acidic foods.

Many of the white women subjects believed AIDS was accompanied by fevers, transmitted through concentrations of impurities and germs, and associated with venereal disease. They recommended cleaning the system with lots of fluids, especially fruit juices. Sweating out the germs associated with AIDS could be accomplished by drinking hot teas and herbs.

Prevention of AIDS also was related to nutrition and to the notion of building up one's resistance, body strength and blood. In addition to foods, moderation in rest, exercise, smoking, alcohol intake and not getting chilled were recommended to build up resistance. Among African American women, foods that were considered to build up one's strength and blood against AIDS were red meat, liver, red wine, grape juice, egg yolks, nutmeg, vitamins, pepper and chocolate. However, these were not to be eaten as a steady diet or they would cause other illnesses. Rather they were to be eaten when "the system is low" or the "body not at its highest peak". Persons who were too thin, old, young, pregnant or menstruating might have to strengthen their resistance to the virus.

White women also believed that if a person's immunity was "run down" or if one had "low blood deficiencies" he or she would be more susceptible to AIDS. Red meat and vitamins were recommended to "build up defenses" or "build up the immune system". Latina women recommended maintaining balance in the diet as a method of avoiding AIDS and emphasized the need to purify the body and blood with foods, herbs and purgatives on a regular basis to prevent AIDS.

Some of the women in each of the ethnic groups had had experience with persons with AIDS. These women often equated AIDS with pneumonia. If the pneumonia was not accompanied by fever and blood, it was usually considered a cold illness by them and treated with medicines and foods with hot properties or blood building properties. For these women, food was thought of and used as nourishment to the body, to assist in convalescence and recovery and to strengthen resistance. Warm soups and liquids were used early in recovery from AIDS pneumonia. As persons with AIDS regained strength and were able to tolerate heavier foods after an episode of pneumonia, blood builders such as meat, eggs, liver and vitamins were recommended by African American subjects.

Latina women used warm foods during the recovery period after AIDS pneumonia because these are more easily digested and because warm foods rebuild the blood and body strength. Many foods considered "neutral" or warm by the subjects were recommended in small amounts for the beginning of the recovery period: flour tortillas, rice, bread, potatoes, frijoles and ginger tea. As the person regained strength, meat, eggs and a little wine were added to the diet to help the body regain health. Vitamins and iron were also considered essential to rebuilding strength. White women who had had experience with AIDS pneumonia recommended warm, bland and soft foods for the recovery period: chicken soup and other soups, tea, toast or bread, oatmeal or cooked cereal and soft boiled or scrambled eggs. These foods were identified as "easy on the stomach". This description and the foods recommended seemed similar to the beliefs of black and Latina women about the diet during the recovery period from AIDS pneumonia.

COMPATIBILITY OF FOOD THERAPIES AND TREATMENT FOR AIDS

Some traditional beliefs held by the white, black, and Latina women in these studies[7-9] about the use of food as therapy would be considered compatible with the biomedical treatment of the AIDS indicator diseases (e.g. opportunistic infections, neoplasms, wasting and so forth). Others, however, would be detrimental to treatment and the person's health. Many of the opportunistic infections associated with AIDS are accompanied by fever. The use of increased amounts of fluids in the diet would not be incompatible with the treatment of fever. On the other hand, the notion of "flushing out the system" or "sweating it out" especially when herbs, minerals oils, laxatives and purgatives are used to accomplish this end would be considered risky. This is particularly true when opportunistic infections are accompanied by diarrhea or when HIV wasting syndrome is the diagnosis. Fluid depletion becomes a real danger to the person with AIDS in these instances. Although diets high in carbohydrates, fat and lactose can exacerbate gastrointestinal problems, high calorie, high protein diets are still recommended for persons with HIV wasting and diarrhea.[27]

The idea of preventing AIDS by building one's resistance through the use of foods has an inherent general health complication in it if the diet becomes too dominated by the so-called "blood builders": red meat, egg yolks, liver, chocolate and so forth. Such a diet might be considered cholesterol rich in current biomedical thought. However, most of the respondents recommended this diet with a view toward moderation in its use. None of them believed it should be a steady diet and they all cautioned that this diet should be used when the "system is low". From an AIDS perspective, the "blood building" diet would have to be used cautiously in order to prevent polymicrobial enteric infections which are potent immunosuppressants. To reduce microbes in the diet, meats must be well cooked, and prepared foods such as egg salads and cold meats should be avoided.[10] The "blood building" diet might contribute to these infections if blood products are not cooked well and prepared foods are not properly refrigerated or are held too long.

The womens' use of bland, soft, warm foods for the recovery period from AIDS-related pneumonia might be considered a beneficial use of traditional food therapy. Generally, persons with *Pneumocystis carinii* pneumonia (PCP) are allowed to eat whatever they can tolerate if these foods are not contraindicated by their medications.[27] Several of the current medications in use for PCP produce adverse side effects such as anorexia, vomiting, diarrhea, alteration in taste and hyper- or hypoglycemia. All of these side effects would have to be considered in developing a nourishing but tolerable diet for persons with PCP.

IMPLICATIONS FOR THE HEALTH PROFESSIONALS

Health care professionals involved with persons with AIDS should have a working knowledge of lay conceptualizations of illness and food therapies. Framing prevention and treatment approaches within lay conceptualizations of health, illness, AIDS and the therapeutic use of food may well have a greater chance of success than insisting on a biomedical approach. Harwood[14] makes the point to health care workers that the probability of changing an individual's conception of disease and treatment in a few short encounters is very small. This is especially true when that conceptualization orders a great deal of health behavior, is supported by the family and social group and is frequently validated by the prescriptions of the biomedical system. It is far more productive, in a pragmatic sense, to accept and work within the existing system of beliefs than to simply impose modern medical regimens which may not be adhered to.[17]

Health professionals should construct therapeutic dietary regimens which are known to be clinically effective and then reconfigure these, to the extent possible, to be compatible with the client's dietary beliefs. When counseling a

client, the approach of the health care worker would be, first, to try to catego-
rize the client's nutrition beliefs and practices as beneficial, neutral or harmful
in the context of an AIDS diagnosis and the various AIDS-related infections
that may occur.[2,14,17] Nutrition counseling within a lay conceptualization of
illness and diet therapy should be guided by four principles: (1) reinforcing
beneficial beliefs and practices; (2) allowing and/or supporting neutral practices; (3) clarifying and correcting harmful beliefs and practices; and (4) adding
information when knowledge is incomplete.[7-9] Whenever possible, added information should be placed within the lay conceptualization of illness and
dietary therapy.

In general, the approach taken to treatment of PCP by the subjects in the
studies by Flaskerud and colleagues[7-9] would be considered beneficial dietary
beliefs and practices and might be reinforced by the health care worker.
Subjects were using bland, warm and soft foods during the recovery period.
Difficulties might arise in taking the medications currently prescribed for PCP
if the person with AIDS were to develop diarrhea as a side effect. To many
ethnic groups, diarrhea is considered a hot illness and antibiotic treatment, also
considered hot, would be thought to be responsible for its occurrence. A way
of dealing with this eventuality would be to neutralize the antibiotic by asking
the client to take it with fruit juice or some other cool substance.[14]

The principle of neutralization is an important one to use in gaining
adherence with a medical regimen when a hot-cold classification is in use. It
should be noted that "neutralization" is an English interpretive term. Spanish-
speaking persons would understand better the use of the term "refresh" in this
context, as in "refreshing" or "cooling" the stomach.

Traditional beliefs about balance or moderation in the diet may be supported by the health care worker when the practices associated with them are
also moderate. For instance, proper diet (a balanced diet) can be supported
because it would have a good fit with the goals of AIDS treatment programs
to enhance immunocompetence. However, drastic measures to correct per-
ceived imbalance (e.g. use of purgatives or eating/drinking blood and blood
products) should be warned against. This warning may take place within the
conceptualization of balance by explaining that when a person is ill, balance
must be restored gradually. Helping the person within his or her conceptual
context would focus on adding greens or fresh fruit to the diet, or well-cooked
meat, slowly and gradually.

The widespread belief in the use of purgatives as a treatment for many
illnesses might be detrimental in the treatment of AIDS. Fevers and upper
respiratory symptoms both may be treated with purgatives to clean out the
system and purify the body. These symptoms are associated with AIDS-related
diseases and might be treated with laxative home remedies in addition to
physician-prescribed treatments. Furthermore, there might be some resistance
to taking antidiarrheal medication because diarrhea is seen as an attempt by the

body to purify itself: to rid itself of toxins. In situations where persons with AIDS are experiencing fevers, upper respiratory symptoms and diarrhea, health care workers should be alert to the possibility that traditional food therapies and health remedies might be in use which are counterproductive or detrimental to treatment. Persons with AIDS and their relatives should be warned within the context of their belief system that these are situations in which the body has to conserve its resources and maintain balance. Too much diarrhea or too many laxatives will deplete resources and upset the balance of the system. The medications and treatments prescribed can be described by health professionals as those that will "build up" the body in these situations.

Dietary education efforts of health care workers must include a recognition that various groups of people will integrate information on AIDS into their already existing belief systems about illness, its treatment and food therapies. Rarely will they create a new conceptualization of illness for new information or a new disease. It becomes necessary to have knowledge of these belief systems in order to provide relevant and effective education, prevention and treatment programs. Dietary education programs may be delivered within the context of the belief system to enhance acceptance. Health professionals may design dietary treatment programs that are congruent with existing beliefs and may be alert to possible conflicts in approach to disease and treatment.

REFERENCES

1. Anderson, J.N. (1983). Health and illness in Philipino immigrants. *The Western Journal of Medicine,* 139(6):811-819.
2. Andrews, M.M. (1989). Culture and nutrition. In Boyle, J.S. and Andrews, M.M. (Eds), *Transcultural Concepts in Nursing.* Glenview, IL: Scott, Foresman, pp. 333-355.
3. Bauwens, E. (1979). Medical beliefs and practices among lower-income Anglos. In Spicer, E. H. (Ed), *Ethnic Medicine in the Southwest.* Tucson, AZ: The University of Arizona Press, pp. 214-270.
4. Berlin, E.A. and Fowkes, W. (1983). A teaching framework for cross-cultural health care: Application in family practice. *The Western Journal of Medicine,* 139:934-938.
5. Cook, C. and Baisden, D. (1986). Ancillary use of folk medicine by patients in primary care clinics in Southwestern West Virginia. *Southern Medical Journal,* 79:1098-1101.
6. Currier, R. (1966). The hot-cold syndrome and symbolic balance in Mexican and Spanish-American folk medicine. *Ethnology,* 5:251-263.
7. Flaskerud, J.H. and Rush, C.E. (1989). AIDS and traditional health beliefs and practices of Black women. *Nursing Research,* 38(4):210-215.
8. Flaskerud, J.H. and Calvillo, E.R. (1991). Beliefs about AIDS, health, and illness among low income Latina women. *Research in Nursing and Health,* 14:431-438.

9. Flaskerud, J.H. and Thompson, J. (1991). Beliefs about AIDS, health, and illness among low income White women. *Nursing Research,* 40(5):266-271.

10. Flaskerud, J.H. (1992). Cofactors of HIV and public health education. In Flaskerud, J.H. and Ungvarski, P.J. (Eds), *HIV/AIDS, A Guide to Nursing Care.* Philadelphia: W.B. Saunders, pp. 314-349.

11. Freimer, N., Echenberg, D., and Kretchmer, N. (1986). Cultural variation: Nutritional and clinical implications. *The Western Journal of Medicine,* 139:928-933.

12. Gilman, S.C., Justice, J., Saepharn, K., and Charles, G. (1992). Use of traditional and modern health services by Laotian refugees. *The Western Journal of Medicine,* 157(3):310-315.

13. Hartog, J. and Hartog, E. (1983). Cultural aspects of health and illness behavior in hospitals. *The Western Journal of Medicine,* 139(6):910-916.

14. Harwood, A. (1971). The hot-cold theory of disease: Implications for treatment of Puerto Rican patients. *Journal of the American Medical Association,* 216(7):1153-1158.

15. Hautman, M.A. and Harrison, J.K. (1982). Health beliefs and practices in a middle-income Anglo-American neighborhood. *Advances in Nursing Science,* 4:49-64.

16. Helman, C.G. (1978). "Feed a cold, starve a fever" — Folk models of infection in an English suburban community, and their relation to medical treatment. *Culture, Medicine and Psychiatry,* 2:107-137.

17. Logan, M. (1973). Humoral medicine in Guatemala and peasant acceptance of modern medicine. *Human Organization,* 32(4):385-395.

18. Jerome, N.W. (1980). Diet and acculturation: The case of Black American immigrants. In Jerome, N.W. (Ed), *Nutrition Anthropology.* Pleasantville, NY: Redgrave Publishing, pp. 275-325.

19. Ludman, E.K. and Newman, J.M. (1984). Yin and yang in the health-related food practices of three Chinese groups. *Journal of Nutrition Education,* 16:3-7.

20. Muecke, M.A. (1983). In search of healers — Southeast Asian refugees in the American health care system. *The Western Journal of Medicine,* 139(6):835-840.

21. Pliskin, K.L. (1992). Dysphoria and somatization in Iranian culture. *The Western Journal of Medicine,* 157(3):295-300.

22. Ramakrishna, J. and Weiss, M.G. (1992). Health, illness, and immigration — East Indians in the United States. *The Western Journal of Medicine,* 157(3):265-275.

23. Roberson, M.H.B. (1987). Home remedies: A cultural study. *Home Healthcare Nurse,* 5(1):35-40.

24. Snow, F. (1974). Folk medical beliefs and their implications for care of patients: A review based on studies among Black Americans. *Annals of Internal Medicine,* 81(1):82-96.

25. Snow, L. (1983). Traditional health beliefs and practices among lower class Black Americans. *The Western Journal of Medicine,* 143:820-823.

26. Snow, L.F. and Johnson, S.M. (1978). Folklore, food, and female reproductive cycle. *Ecology of Food and Nutrition,* 7:47-49.

27. Ungvarski, P. (1992). Clinical manifestations of AIDS. In Flaskerud, J.H. and Ungvarski, P.J. (Eds), *HIV/AIDS, A Guide to Nursing Care.* W.B. Saunders: Philadelphia, pp. 54-145.

28. Wei, L. (1976). Theoretical foundation of Chinese medicine: A modern interpretation. *American Journal of Chinese Medicine,* 4:355-372.
29. Whang, J. (1981). Chinese traditional food therapy. *Journal of the American Dietetic Association,* 78:55-57.

INDEX

A

Ability to inactivate viruses, 125
Abnormalities of glucose metabolism, 218
Accumulate ascorbate to high levels, 121
Acid-soluble thiols (cysteine and glutathione GSH), 126
Action of ascorbic acid on HIV activity, 55
Action of glutathione peroxidase, 120
Activation of NF-×ß, 130
Activity factor, 226
Acute phase, 181
Addictive drugs, 53, 54
  anorexic effects, 50
  effects on
    nutritional status, 50
    food intake, 50, 52
    immunity, 50
    liquid intake, 50
Adipose tissue, 219
Adjuvant, 155
Administration, 206, 209
Adrenal, 220
Adults, 34
Adverse effects on platelet, lymphocyte, fixed macrophage and PMN, 227
Africa, 17, 19
AIDS, 4, 33, 49, 51–52, 53, 168
  children, 35
  epidemiology, 49
  men, 42, 54
  -related complex or asymptomatic infection with HIV, 56
  traveler's diarrhea maybe a life-threatening illness in a HIV infected person, 195
  women, 39
AL-721, 227
Albumin, 54
Amino acids, 170, 172, 226, 228
Anemia, 90–91, 96
Anergy, 55
Anorexia, 177, 219
Anorexigen effect, 52
Anthropometric and immunological parameters, 54
  measurements, 54, 202
  parameters, 55
Anthropometry, 9

Anti-fungal, 224
Anti-oxidants, 13
Anti-retroviral, 224
Anti-retroviral convergent therapies, 221
Antibacterial effects, 124
Antibiotic, 224
Antibodies to cytokines, 217
  anti-HIV, 216
  production, 145
Antioxidants, 24
  depletion, 118
  or free-radical scavenging function, 120
Antiviral, 224
Antiviral action, 125
Apoptosis, 153–154, 219
Arginine, 221
Ascorbic acid, 55, 56
  and thiol effects on HIV replication, 129
  concentrations, 131
  effects on the immune system, 121
  in combination with NAC, 131
  is capable of inhibiting HIV replication, 126
  repletion diet, 127
  Vitamin C, 117, 118
Associated with AIDS, 131
Asymptomatic persons, 127
Attainable in blood plasma, 131
Autoimmune disease, 4
AZT, 176

B

ß-carotene, 14, 93, 131, 144, 150, 152, 153, 218
B lymphocyte rate, 54
B-cell lymphoma, 146
Babies from HIV-infected mothers, 54
Beneficial effects of vitamin C, 122
Beta, 216
Biliary tract disease, 224
Biochemical monitoring, 207
Bioelectric impedance (BIA), 62
Bioimpedance techniques, 9
Biosynthesis of collagen and carnitine, 118
Body
  composition, 217, 218
  fat, 168, 226
  fat mass, 221